Cerecedo Sáenz
Salinas Rodríguez
Hernández Ávila

The Molango Rift

Cerecedo Sáenz
Salinas Rodríguez
Hernández Ávila

The Molango Rift

the Molango Rift and its mineral resources

ScienciaScripts

Imprint

Any brand names and product names mentioned in this book are subject to trademark, brand or patent protection and are trademarks or registered trademarks of their respective holders. The use of brand names, product names, common names, trade names, product descriptions etc. even without a particular marking in this work is in no way to be construed to mean that such names may be regarded as unrestricted in respect of trademark and brand protection legislation and could thus be used by anyone.

Cover image: www.ingimage.com

This book is a translation from the original published under ISBN 978-620-2-15100-9.

Publisher:
Sciencia Scripts
is a trademark of
Dodo Books Indian Ocean Ltd. and OmniScriptum S.R.L publishing group

120 High Road, East Finchley, London, N2 9ED, United Kingdom
Str. Armeneasca 28/1, office 1, Chisinau MD-2012, Republic of Moldova, Europe
Printed at: see last page
ISBN: 978-620-7-66887-8

Contents

Preface

The Molango rift is not only a geological wonder, it is also our planet's past, a silent witness to the terrifying natural forces that acted upon it; it provides us with surprises of the earth's natural treasures with regal minerals, including some of the world's rarest ores, as well as base metals for industry, and it provides us with the earth's energy elements that go from nature to our service.

Although time has obscured the evidence of our planet's primitive epochs, today scientific progress in the areas of earth science has revealed some of its origins, its age, and how life on earth could have arisen.

Thus, the Molango rift is a monument of time, which keeps a calendar of lost worlds. Which are the logbook of the long journey of the continents.

Thus, since the ancient centuries of hypotheses with errors in the formation of the solar system, the evolution of knowledge of the earth and the cosmos has been exponential.

But there are still frontiers and spaces of study to know their relationships; so the natural curiosity of man with the earth leads him to ask ^What is the other world under our feet? How does a rift form? Are rifts the horns of plenty or are rifts completely sterile? Jos minerals are a consequence of rifts? Is it possible to reach such a point of precision to be able to locate mineral treasures in the time and space of rifts? Are the rifts the cornerstones of civilisation because of their strategic metallic contents?

The authors

SUMMARY

A rift is a large fissure in the earth's crust, which starts as a small crack derived from stresses generated in the tectonic plates, and when interacting with each other mainly in subduction zones, they cause a geodynamic movement within the crust. Therefore, over millions of years in the geological history of a rift, different environments can be associated with different types of mineralisation during or after the rift sequence, and due to the phenomena associated with the stresses generated during the formation of the rift.

For example, the formation of stratiform and vein deposits is attributed to the precipitation of minerals from hydrothermal solutions. In certain cases, reconstruction of the tectonic regime, lithological association and chemical characteristics, including REE content, allow the correlation of ancient sedimentary deposits with those currently forming on the seafloor or in continental hot springs.

Here we present a case study of the prospection of a Sedimentary Exhalational Exhalative Deposit (SEDEX) in the manganese-ferrous district of Molango Hidalgo, for this purpose we carried out a stratigraphic, tectonic and geodynamic analysis of the study area allowing us to know the location of the basin associated with the location of a deposit of these characteristics. In this work we obtained evidence that allows us to determine that the deposit is located in a siliciclastic sequence of reducing environment, the tectonics is compatible with an extensional geodynamic model. In a sequence with high silicification and that presented two types of mineralization in a sedimentary sequence, the minerals presented by X-ray diffraction (XRD) were: pyrite, chalcopyrite, pyrrhotite, sphalerite, alpha quartz and high quartz.

The Induction Coupled Plasma Spectrometry (ICP) analysis of the representative samples of the outcropping sandstones and shales of the Huayacocotla Formation of the undifferentiated member studied in this paper 6 presented base metal values of 30 ppm Zn and 9 ppm Cu. Similarly high elemental values were obtained with the following values 82 ppm for Ba, 0.9 % Al, 0.17 % Ca, 1.64 % Fe, 0.08 % Ti, 40.8 % Si, Bi 2 ppm , 20 ppm Ce, 2.2 ppm Co , 30 ppm Cr, Cs 2.7 ppm , 0.9 ppm of Er , Ga 2.5 ppm , 1.6 ppm Gd , Ge 1.5 ppm , 9 ppm of La, 71 ppm Li , 104 ppm Mn, 10 ppm Nd , 17 ppm Rb, 2ppm Se, Sr 9 ppm , 10 ppm Ta , Te 6 ppm , 1.3 ppm of U , V 28 ppm , 9 ppm , and 0.7 ppm Yb. While in lower proportion of precious metals were found as Au < 0.02 g / t, Pd <0.05 g / t, Pt <0.05 g / t.

On the other hand, as a result of the exhalative rafts in the exhalite minerals in an extensional geodynamic environment, we were able to classify this SEDEX deposit as a Shelwyn subtype.

INTRODUCTION

Exploration work is currently underway to locate mineral deposits of economic interest. The study of geological structures, their formation processes and their influence on the enrichment of strategic and precious elements for the development of new and improved technological applications is therefore of importance for the world economy.

Similarly, there is much information linking the mineralisation of different deposits to the evolution of particular geological structures.

In particular, in the case of the Pb-Zn mineralisation associated with the Jurassic age in the Mor Range in Pakistan, which was emplaced during the formation of the range (syngenetic), as well as at a later stage (epigenetic), being of SEDEX Type [1], with clayey limestones as stratiform replacement, pyrites in sphalerite matrix with galena in black shale. Seam-like fissure fillings, composed of Lower Jurassic carbonates and a sequence of marine siliciclastic rocks. In some cases there is evidence of remobilisation which helps to establish a synsedimentary character, where the grain size of the minerals contained is usually considerably fine [2, 3].

Furthermore, in supergene Zn non-sulphide deposits, Fe oxyhydroxides are mainly concentrated in residual zones (Gossan) [4], which form in systems containing elements such as Zn, Pb, Co, REE, Sc, Ga, Ge, V, among others. Similarly, Cu-U-V deposits were emplaced in the same tectonic setting, hosted in Triassic red-bedded sandstones and located at the Eureka mine in northern Spain [5], showing supergene alteration with nickel and cobalt dispersion. Vanadium was found as vanadates, while bismuth as oxides and rare earth phosphates as monazite and xenotime. This is important, because it is widely 8 accepted that Red Bed sedimentary deposits and other stratiform sedimentary deposits occur in such environments [6].

Meanwhile, an outcrop showing Nb-Ta mineralisation [7], present in a bimodal sequence of peralkaline felsic rocks, is described in Mongolia. A1И, tectonic processes produced mineralization of Zr, Nb, heavy rare earths, Y, U, Th and Ta; while peralkaline felsic rocks formed Li-F with mineralization of Sn, W, Ta, Li and Nd. In addition, the occurrence of peralkaline granites is present in northeastern Mongolia, where Nd-Ta is associated with REE and zircon mineralisation.

Similarly, it has been mentioned that other alternative sources for locating rare earths are sediments from the ocean floor, the continental shelf, rivers, streams and lakes. There are also deposits of phosphorites, industrial wastes such as red mud, phosphogypsum, coal fly ash, mining waste, acid mine drainage and electronic waste [8].

In particular, a rift is a large fissure in the Earth's crust, which starts as a small fissure derived from the stresses generated in the tectonic plates, and as they interact with each other mainly in subduction zones, causes a geodynamic movement within the crust. Therefore, over millions of years in the geological history of a rift, it is possible that several types of mineralisation can be associated during or after the rift sequence, and due to the phenomena associated with the stresses generated during the formation of the rift.

Therefore, in the tectonic framework of different Rifts explored around the world, different types of mineralisation have been found. For example, at Tongkuangyu in northern China, Cu mineralisation of Paleo-Protezoic age is described due to two events; the first is red-layer copper sulphides that formed during metamorphic regression, and the second is vein-type sulphides derived from the remobilisation of first-stage sulphides by infiltration of external fluids, such as residual seawater and metamorphic fluids at the surface [9].

Similarly, the gold mineralisation linked to the northern China Haoyaoerhudong Grabo is related to a Rift-like extensional structure, with high gold contents (148 g/ton) hosted in protezoic strata generated in a post-orogenic extensional environment. This same Gabbro has calc-alkaline characteristics, enrichment of large ion lithophile elements (LILE), as well as light rare earth elements [10].

On the other hand, one of the largest Fe-REE-Nd deposits in the world is Bayan Obo in Mongolia, which is located in a rift-like structure [11]. Similarly, it was determined that the fluids that form this mineralisation are similar to deposits located in various parts of the world, which contain carbonatites and alkaline ^gneiss rocks with adequate REE contents. In addition, other elements with Fe oxides and Cu-Au-REE contents have a different fluid source in the associated ^gneiss rocks found in the Bayan Obo deposit. Also, outcrops with good gold and Fe contents [10] have been found in some rift-like structures in hydrothermal magmatic systems.

Thus, the occurrence of REE associated with rift-like structures has evidence worldwide. Exploration work is currently underway to locate mineral deposits of economic interest. Therefore, the study of geological structures, their formation processes and their influence on the enrichment of strategic and precious elements for the development of new and improved technological applications is of importance for the world economy.

Currently, this type of prospecting has not been carried out in Mexico, so it is interesting to carry out these pioneering studies like the one presented here, as they could provide valuable data on the presence of strategic minerals (REE) and those of the platinum group (PGE).

In this way, this work contributes to the innovative consolidation of the exploration of strategic mineral deposits, as it has allowed the validation of areas of possible exploitation on the continent, while minimising the direct costs of exploration.

Finally, the results shown are both effective and cost-effective in determining the possible geodynamic environment for the classification parameters of rift-like structures, starting with the field recognition of strategic correlations related to marine transgressions contemporaneous with rifting in areas that are currently emerging.

BACKGROUND

For the development of this work, an indirect prospecting method was used, which was extensively described in a platinum and gold prospecting work in this area [18]. According to that work, firstly, fieldwork was carried out looking for the base of the transgressive stage that was previously identified as a marine siliciclastic sequence of Lower Jurassic age. Also, the information obtained was correlated with other outcrops such as the one found in the San Cayetano Formation in Matahambres, Cuba [19]. On the other hand, in order to properly identify the outcrop and perform the sampling, it was considered to analyse the marine transgressions and to consider in particular, the formations that contain a geological sedimentary deposit of lower Jurassic age [20], integrating geological, stratigraphic and structural studies. Thus, when locating the outcrop, samples were obtained by drilling in the uncovered area. The drilling was carried out with a portable drilling rig (JKS, Winkei, GW-15), with an air-cooled, gas-driven, two-cycle motor. The drill core obtained was at a depth of 9 metres and 0.03175 metres in diameter. The as^ obtained samples were homogenised and cut into quarters to obtain a single homogeneous sample to be characterised. The characterisation carried out in this work was executed in order to obtain accurate data of the mineralogical phases present, for which a general phase analysis by X-ray diffraction (XRD) was proposed, where the samples were crushed to an average particle size of less than 78 pm and put into an Equinox 2000 X-ray diffractometer (INEL, Artenary, France, located at UAEH) with CoKa 1 radiation. Phase identification was based on COD Inorganics 2015 databases in Crystallography Open Database Math Software, (v.1.10, Crystal Impact, Bonn, Germany). In addition, characterisation studies were completed by Scanning Electron Microscopy (SEM) to identify texture, granulometry and morphology of the detected stages; Similarly, analysis by X-ray mapping, and Energy Dispersive Spectrometry (EDS) helped to determine the point and semi-quantitative composition of some of the previously identified phases, using a JEOL scanning electron microscope model JSM-IT300 (JEOL Ltd, Tokyo, Japan, located at UAEH) and an OXFORD X-Ray detector (OXFORD Instruments, Oxford shire, UK) with an accelerating voltage of 30 kV. To obtain a representative analysis, the powder samples were placed on the sample in uniform layers and quantitative spot routines were run over large scan areas (about 4.5 mm 2).

For chemical analysis, an inductively coupled plasma spectrometry (ICP-MS) analysis was performed by Actlabs (Activation Laboratories Ltd., Ontario, Canada) to find the average total rock composition of the mineralised phase, where positive anomaHas light rare earth and platinum group mineral (PGE) contents. The methodology used in this case was as follows; samples were fused and then diluted and analysed by a Perkin Elmer Sciex ELAN 9000 ICP- MS spectrometer (Located at Actlabs, Canada). The fused blank was run in triplicate and duplicates were run every 10 samples, then the instrument was recalibrated every 44 samples. Finally, to determine the Au and Pt contents of the ore concentrates, a cupellation analysis was carried out on an EMISION CL series furnace. Sample preparation was done using borax, PbO, bone ash, sodium carbonate as a flux, as well as silver (99.99 purity); the melting temperature was 1000 °C for 90 minutes. Afterwards, the slag separation process was carried out, obtaining a button containing Ag, Pt and Au values. Then, the release of Au from the button was performed in porcelain crucibles on a hot plate at 120 °C, adding 15 % nitric acid to dissolve Ag, obtaining a silver nitrate solution. Then Au and Pt release was performed using aqua regia with the addition of 10 % chlor^dic acid with stirring in test tubes. The determination of Au and Pt contents was done using a Varian 735ES ICP-OES ICP, where samples were analysed with a minimum of 10 certified reference materials, all prepared with sodium peroxide fusion. Every 10 samples were prepared and analysed in duplicate and the blank was renewed every 30 samples. Internal standards were used as part of standard operations.

On the other hand, there are two aspects to consider; the classification of the deposit and the mineral association. In the first case, it was considered to be a SEDEX-type mineralisation because two types of mineralisation were recognised: phylonic and stratiform as previously reported [19]. In the phylonic outburst, rates were observed in the stockwork towards the base, which of course did not cut the upper part of the sedimentary sequence. Here, it is inferred that they may correspond to emission channels, as pointed out by Wang et al [21]. On the other hand, the SEDEX [1] type mineralisation is stratiform and concordant with shales, shale sediments of submarine origin, as well as in some areas, presenting great schistosity [22] and faults [23]. Therefore, it could be inferred that the age of this mineralisation predates the development of the mountains and hills in this area, which is associated with schistosity and pyrite. Additionally, it is inferred here that, perhaps due to reductive conditions, the precipitation of the sulphides observed in the hand sample was favoured, such as finely disseminated pyrite [20] and sphaletine in lesser proportion [24], which in some cases, shows framboidal crystallization [25] and botroidal growths, showing that its formation corresponds to a chemical nature, formed in a submarine environment [2,3]. Thus, all this approximates to a Selwyn-type SEDEX mineralization [26], because the nature of the Lower Jurassic-age Huayacocotla Formation is of a siliciclastic nature of a SEDEX-type mineral reducing environment [1, 13]. Finally, the low amount of nickel and other metallic contents such as chromium and the association of Fe and Mg [27]. Similarly, other elements of ultrabasic affinity were present in this mineralisation, and the additional results found allow classifying this site as SEDEX [12, 13]. However, considering these results and the significant platinum contents in metallic form, this deposit can be subclassified as a more regional subtype.

1.1 Geological framework.

In terms of regional geology, the works of Cantu-Chapa [28,29,30], who, based on biogeographical evidence from Permian to Lower Jurassic ammonites, provide solid evidence to explain the origin of the Gulf of Mexico with close affinity to the western margin of Pangaea. He describes three keys to this; during the Bajocian he mentions that there is evidence for the occurrence of the ammonite stephanoceras during the western margin of America (Alaska, Canada, USA, Venezuela, Peru, Argentina and Chile) and here in Mexico in Oaxaca.

The second in the Batonian - Calovian Wagnericeras ammonite appears in a transgressive cycle in eastern and southeastern Mexico. This event is particularly important because it very precisely locates the emplacement of the SEDEX mineralization studied here. And the third clue is provided by the occurrence of many known Permian to Jurassic cephalopods that are relative only to the margin of the

Pacific province.

Likewise, Carrillo [31], carried out a detailed geological study of the Calovian red-bedded continental Formation called the Cahuasas Formation, formed by powerful thicknesses up to two thousand metres; particularly this Formation is important for the purposes of this work because it represents a change in the sedimentation rate, where in a relatively short period during the Calovian a great thickness of clastic sediments was deposited. It thus confirms a criterion for classifying an ancient rift-type megastructure.

For this work, the lithological description of Ochoa [32] of the Molango Hidalgo manganese-^ferrous district was also taken into account, enriching this information with other authors who have studied the area. The outcropping formations are mentioned in more detail below.

LOCAL GEOLOGY

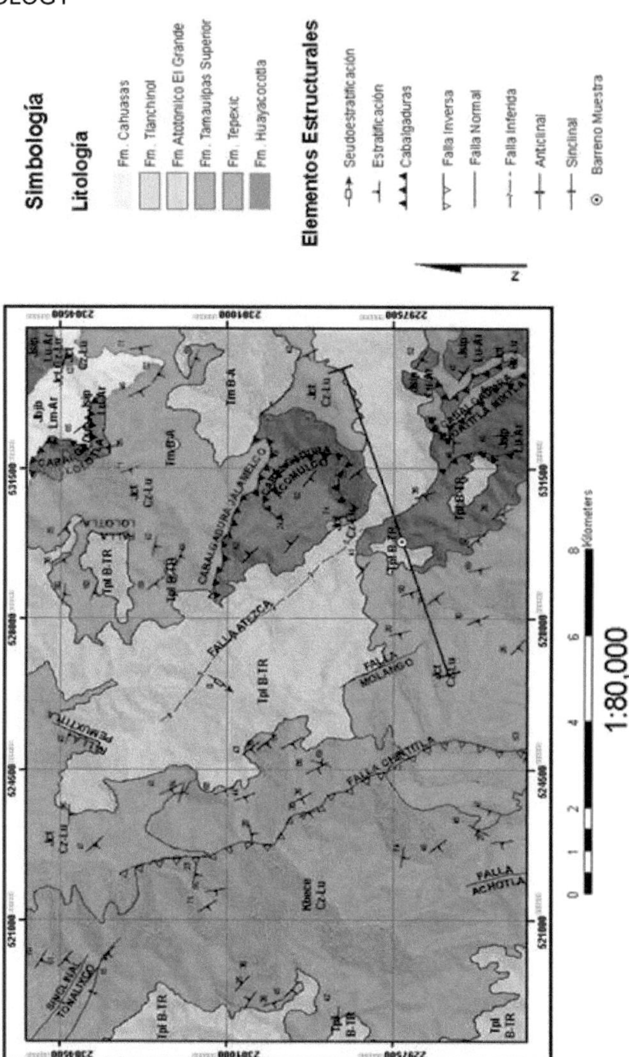

Figure 1. Geological map of study area 1:80,000

1.2 Stratigraphy
Precambrico Gneis
Huiznopala

This Formation was first described by Von Kuegelgen in Ochoa [32] while its name was assigned by Fries [34], its type locality is in the valleys of the Rfo Claro, the Agua de Sal creek and the Acatepec creek and are the oldest rocks in the study area which were described as two classes of metamorphic crystalline rocks, orthogneiss and paragneiss.

Lower Jurassic
Formation Huayacocotla

The Huayacocotla Formation is defined by Erben [33]. In general, it consists of dark shales with intercalations of sandstones and conglomerates with sparse limestone lenses, has considerable vegetal material in the upper and lower part of the top and contains carbon towards the Imlay top [34]. In Hidalgo its maximum thickness is exposed; the lower member consists of conglomerate, sandstone, siltstone and shale with exoclasts, which contain reworked fossils such as fusilinids and crinoids, the intermediate member is made up of conglomerate, sandstone, siltstone and shale with ammonites and the upper member is formed by sandstone, siltstone, shale and conglomerate with the presence of continental fossil plants. The thickness of this formation varies between 500-1000m, with a progressive thinning of the southern outcrops towards the north and east. Its lower and upper contacts are generally discordant with the Cahuasasas Formation; in some areas where the Cahuasas Formation loses its continuity, the Huayacocotla Formation can be found covered by the Tepexic Formation in an angular unconformity Ochoa *op.cit* [32]. On the southwestern and northeastern flanks of the rift.

It is worth mentioning that further north 300 km in the Huizachal-Peregrina Anticlinorium, the stratigraphic equivalent of the Huayacocotla Formation is called the Aloformación la Boca Rueda-Gaxiola [32] who based on a palynomorph assemblage found in this alloformation, Rueda *op. cit.* interprets a coastal environment, in a poorly circulating basin. Relatively more recent interpretations consider that the presence of a relatively varied fauna of ammonites and peletipods suggests low-energy shallow marine depositional conditions in shelf areas close to the mainland.

Figure 2. In 1, outcrop of the middle portion of the Huayacocotla Formation at the Rio Chinameca bridge; in 2, a normal fault in the Huayacocotla Formation; in 3, the upper

unit of the Huayacocotla Formation of shale with plants.

Middle Jurassic

Formation Cahuasas.

These rocks were originally considered to be part of the Huizachal Formation [31,32]. Carrillo [31] informally proposed the name Cahuasas for the red beds outcropping in the Huayacocotla Anticlinorium of Middle Jurassic age. Later, in 1965, the same author formalized this proposal. The type locality is located at Rancho Cahuasas, Hidalgo, on the Rfo Amajac, southeast of Chapulhuacan, Hidalgo. In general, this unit is made up of red sandstone, conglomerate and siltstone; it also adds that on the Tianguistengo-Rancho Mixtla road, the unit is represented by approximately 4 m of poorly sorted conglomerate in thick layers, composed of subangular fragments of red to dark grey quartz sandstone; the diameter of the fragments varies from 1 to 15 cm approximately; immediately above is a 35 m package of red siltstone and clayey sandstone, in medium to thick layers; on the other hand, the Rfo Amaxac ravine is composed of poorly sorted conglomerates, sandstone, conglomeratic sandstone, shale and red siltstone containing abundant white mica flakes; the sandstones and conglomerates are frequently cross-stratified. Carrillo *op.cit.* reports more than 1200 m in the Rfo Amajac ravine; while in the Rfo Claro it is represented by more than 250 m. Around Tianguistengo, Hidalgo, the Cahuasas Formation discordantly underlies the Taman Formation and discordantly overlies the Huayacocotla Formation; in the Rfo Amajac, this unit overlies limestones and sandy limestones with Callovian fauna and overlies sediments of the Huayacocotla Formation; in the Rfo Claro south-southeast of Huayacocotla, Veracruz and in the Huehuetla-Cueva Ahumada region, it also rests on Liasic (Lower Jurassic) layers and is discordantly overlain by Upper Jurassic limestones [32]. Salvador in Maynard [27] mentions that the lithologic composition, the absence of fossils, the geographic distribution, the abrupt change in thickness and the stratigraphic relationships suggest that the Middle Jurassic red beds were accumulated as alluvial fans, and as fluvial and lacustrine deposits; he also adds that the presence of non-marine red beds overlying the Huayacocotla Formation (marine) is indicative of a regression during the early part of the Middle Jurassic. Because of its lithological similarity with the La Joya Formation, it is inferred that this unit was deposited on alluvial fans, on a continental alluvial plain above the supramarean line in arid or semi-arid conditions; on a marine alluvial plain below the supramarean Hmite or by fluvial flooding. On the other hand, Salvador [38] tentatively correlates it with the La Joya Formation (Huizachal-Peregrina Anticlinorium), based on lithological similarity and stratigraphic relationships. Carrillo-Bravo [31] based on its stratigraphic position assigns it a younger age than the Pliensbachian and older than the Callovian; i.e., it spans part or perhaps all of the Middle Jurassic.

Figure 3, Different lithologies of the Cahuasas Formation. In 1, the Cahuasas Formation, with red beds of poorly sorted clasts from 1 cm to 50 cm, in figure 2 fine lithology of marly aspect, in figure 3 normal fault with chloritic alteration, figure 4 intraformational fault with quartz and barite, figure 5 copper in red beds.

Formation Tepexic.

The first mention of this unit was made by Imlay [34]. It is located at the Mamposteria de Acazapa bridge, downstream of the Tepexic plant. Erben [33] mentions that this unit is constituted by impure grey to dark grey, coarse-grained limestones, with many quartz grains; it also presents dark grey to almost blackish grey calcarenite; the stratification is poorly marked, the banks reaching thicknesses of approximately 30-50 cm; in many occasions the base of the sequence is formed by a conglomerate limestone. The lower Hmite of the Tepexic Formation is concordant with the Huizachal Formation; while in other areas it rests in discordance on the Cahuasasas Cantu-Chapa, [28,29,30], Huayacocotla, Suter in Okita [20] formations. The uppermost Hmite is generally concordant with the formation; however, it has been documented that in some areas this unit underlies the Taman Formation and the Chipoco Formation. Finally, Cantu-op.*cit.* mentions that northwest of Poza Rica this unit rests on the Palo Blanco Formation. Erben mentions that this unit represents transgressive marine deposits and that the fauna found (pelarthropods and ammonites) indicate that it was deposited near the coast, or in the somewhat deeper parts of the coast. On the other hand, Suter [20] mentions that the deposition of this unit initiated a marine transgression and that it was deposited over an elevated area, which coincides with a part of what was called Alto de Ixtla for the depositional period of the Cahuasasas Formation; he also mentions that based on paleontological data reported by Aguayo [35] and Cantu-Chapa [28] the depositional environment changed from littoral to deeper over time. In the original designation, Erben [33] proposes a Callovian age for this unit. In the Sierra Madre Oriental Cantu-Chapa [30] assigns an age of middle Callovian, based on the presence of the ostreid *Liogryphaea nebrascensis* and the ammonites of the genera *Neuqueniceras* and *Reinekeia*. In the Poza Rica subsoil, Cantu-Chapa [30] interprets this unit as having a stratigraphic range from the Batonian to the Callovian. Cantu-Chapa [29,30] mentions that the Santiago and Huehuetepec Formation (respectively) represent a lateral equivalent of the Tepexic Formation. Humphrey and Imlay [34] consider this unit to be equivalent to the Minas Viejas Formation of the Gulf of Sabinas in NE Mexico, and to parts of the Novillo-Zuloaga Formation.

Figura 4.   Tepexic Formation outcrop

Upper Jurassic
Formacion Santiago.
This unit was proposed by Cantu-Chapa, [28], on the western slope of the Rio Moctezuma (Taman, San Luis Potos^, near the mouth of the Arroyo Santiago, from which the name of this formation is derived. At this locality, the section presents a certain structural complexity, so it is considered convenient to propose a main reference section (lectotype) or several reference sections, the latter would constitute a stratotype composed of "dark grey calcareous shales up to 40 cm thick, weathering to brownish-brown or reddish-brown, they present a diagonal cleavage to the bedding planes which sometimes leads to misjudgements in the measurement of the structural data, and they also have intercalated calcareous nodules". Cantu-Chapa [29] recognizes that the base of the formation in Pisaflores, Tajo de Tetzintla, Hidalgo; Huauchinango-Villa Juarez, Puebla, is represented by black to dark grey, highly fractured shales with dark grey calcareous nodules and in Huehuetla, Hidalgo by intercalations of black, carbonaceous shales and limestone layers varying in thickness from 2-3 m and 5-10 cm respectively. In his type locality Cantu-Chapa [28] measured a thickness of 160 m, he mentions the presence of asymmetric folds that

16

could imply repetitions in some points of the section, which would have repercussions on the real thickness. The lower Hmite is transitional and concordant with the Tepexic Formation, except in the basement of the Soledad-Miquetla area, where it overlies the Palo Blanco Cantu-Chapa Formation, [28] and in the Huiznopala area, where it rests directly on the Huiznopala Ochoa-Camarillo Gneiss [32]. The uppermost Hmite is transitional and concordant with the Taman Formation except at Molando, Hidalgo, where it discordantly underlies the Chipoco Formation. On the other hand, at Taman, San Luis Potosi Pisaflores, Hidalgo and Rfo Tezcapa, Puebla, the lower Hmite does not outcrop. In the middle part of the Santiago Formation, it was deposited in quiet marine reducing conditions in basin facies. Cantu-Chapa [28] assigns an age of middle Callovian - upper Oxfordian to this formation. In the localities of Pisaflores, Hidalgo and the Huauchinango-Villa-Juarez highway, Puebla, he assigns an age of middle Callovian based on the presence of ammonite species belonging to the genus *Reineckeia*, in these localities only the lower part of this formation outcrops. Cantu-Chapa records the presence of the ammonite genera *Perisphinctes* and *Dichotomosphinctes* at the type locality and mentions the faunal similarity to that found at San Pedro del Gallo, Durango, which allows him to assign a late Oxfordian age. Cantu-Chapa mentions that the Tepexic Formation, besides being older than the Santiago Formation in some localities, also represents a lateral change with the Santiago Formation during the Middle Callovian, the latter based on the similarity of the palaeontological content between the two formations.

Figura 5.   Clay lithology of the Santiago Formation

Formation Chipoco

It was defined by Hermoso de la Torre [35], The type locality is in the area of the Tetzintla pit, located next to the town of Chipoco. As a set of sedimentary rocks arranged with an alternation of limestones and dark grey calcareous shales. The age has been assigned to this formation based on *Idoceran* and *Gloceran ammonites* comprising the Early Kimmeridgian to Lower Titonian Ochoa [32].

Figura 6.   Outcrop of the Chipoco Formation

Pepper Formation

Defined by Heim, in Ochoa [32]. At Rancho Pimienta, located approximately 300 m west of the Mexico - Laredo highway (Km 337 - 338), in the territory of San Luis PotosL. It consists of limestone with a texture of "mudstones of light colours, clean or with little clay, planktonic microforaminifers and lenses of flint; black mudstones, recrystallized, dark brown; sometimes with abundant Saccocoma sp. in the middle part and clayey limestone and with radiolarians in the lower part. This formation concordantly underlies the Chapulhuacan Formation, east of the Valles-San Luis Potos^ platform and the El Abra Formation. In the Valle de Guadalupe well (PEMEX), it concordantly overlies the Taman Formation. The age of this formation was determined on the basis of its ammonite content to be Tithonian and is correlatable in time with the La Casita and La Caja Formations.

Its depositional environment is an unstable submerged platform, with calm, clear waters and normal salinity. It varies from outer shelf to basin, with low energy (PEMEX, 1988, Salvador [36].

Figura 7.   Pepper Formation Outcrop

Formation Tlanchinol

This Formation was nomenclature by Robin [37], for volcanic rocks outcropping along the road from Tlanchinol to Huejutla, Hidalgo. The type locality is not defined, however, Suter [20] mentions that the rocks outcropping along the Carretera Federal 105, north of Tlanchinol can be considered as the type section. Ochoa [32] mentions that at Cerro Las Puentes, Hidalgo, this unit consists of a series of basaltic outcrops intercalated with pyroclastic horizons, airborne tuffs and some andesphitic outcrops. The thickness reported for this unit is very varied; stepwise melt thicknesses of approximately 30 m have been reported north of Tlanchinol up to a maximum thickness of 750 m observed at Tlanchinol and Quetzalzongo. This unit rests on Precambrian rocks, Jurassic and Cretaceous marine sediments, with marked angular and erosional unconformity; while it rests discordantly on Huiznopala Gneiss and Jurassic and Cretaceous sequence formations and underlies rocks of the Atotonilco El Grande Formation. The Tlanchinol Formation is discordantly underlain by the Chicontepec Formation in the towns of Olotla and Tlamamala, Hidalgo State. The alkaline basalts of this formation are considered to have been produced by fissural volcanism during the Upper Miocene.

Formation Atotonilco El Grande

This formation was defined by Geyne [37] and is found in the periphery of the town of Atotonilco El Grande, State of Hidalgo, The type locality of this formation is located in the periphery of the town of Atotonilco el Grande, State of Hidalgo. It is mainly composed of layers of variable texture from mudstone to conglomerate. In the area the lithology consists mainly of a sequence of bimodal volcanic volcanism of volcanic rocks such as basalts, intercalated with tuffs of rioHtic, dacitic and andesphitic composition; lindinzite since we were in the centre of what was once a volcano and in the part of the escarpment, in the lower part there are blocks of extension indicating the symmetry of a rift, in the same way Robin [37] considered these tuffs as ignimbrites. And the thickness of this Formation is estimated to be more than 400 m, according to outcrops in the vicinity of Molango. The rocks of the Atotonilco El Grande Formation at a PHocene age.

Figura 8.   Basalt spills from the Atotonilco El Grande Formation

1.3 Transgressions in the Molango area, Hidalgo.

The most important work modelling the direction of transgressions in Mexico is that of Cantu Chapa, [38]. Figure 2, ah shows the zones of the main marine transgression in the Lower Jurassic, which shows the Huayacocotla Formation. During the Liasic, the possible persistence of Triassic brackish seas took place in the central part of the Republic, especially in the states of Hidalgo, western parts of Veracruz and northern Puebla, Cantu [30]. The lower portion of the Lower Jurassic, (Huayacocotla Formation), is siliciclastic marine by deposition of a marine incursion from the Tethys, and possible transgressions and regressions during the Lower Jurassic are considered. The accumulation of these sequences is dominated by continental rocks (La Joya Formation, Nazas Formation) due to the evidence of plants in the latter, such as Otozamites, Ptilophylum and Williamsonia, from a coastal environment. As shown in Figure 2, possible transgressions and regressions during the Lower Jurassic are considered in most of the emerged terrain of the republic.

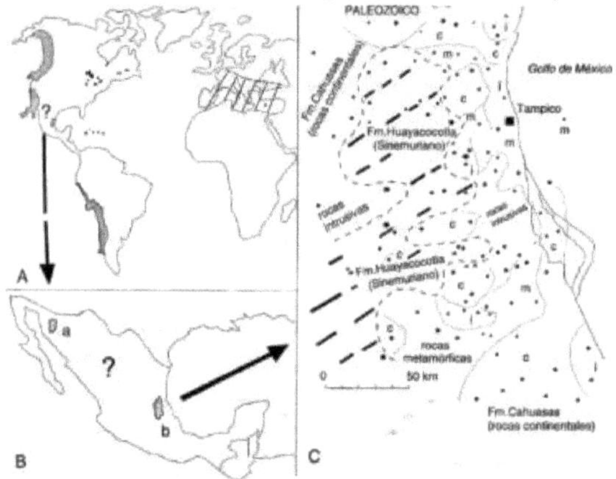

Figura 9.   Jurassic marine transgressions in Mexico.

The Gulf of Mexico during the Middle Jurassic had an abrupt change in the thickness of its basement, with accumulations of red beds (Cahuasas Formation) prograding into alluvial fans and fluvial and lacustrine deposits of the rift basement or degraben system, where the stage of the beginning of the break-up of Pangaea gave rise to the Gulf of Mexico, generating the North American plate, the South American plate and the African plate [36]. During the Middle Jurassic there was an emersion of the Mexican continent, with evidence of initial uplift from the Early Jurassic (La Joya Formation, La Boca Formation) [31], in the shallow waters of the gulf with high and low energy conditions deposited calcarenite and shale beds (Tepexic Formation, Santiago Formation), the irregular layering of calcarenite and shale beds (Tepexic Formation, Santiago Formation), the irregular arrangement of the deposits indicate the existence of islands and channels in the sea in the east-central part of Mexico during the Calovian and the break-up and separation of Pangaea that led to the formation of tectonic pillars and tectonic trenches. An extensive transgression that began in the early Late Jurassic and culminated in the Late Cretaceous caused the sea to migrate into graben areas delineating islands and peninsulas that control sedimentation patterns. Meanwhile in central Mexico during this age interval the Santiago Formation was deposited, consisting of black, microcrystalline, well-stratified, fine-grained limestone and black shale of the Cantu-Chapa open shelf environment [30].

The major transgression, which is represented by black limestone facies with rhythmic intercalations of black shale, has nodules, bands and lenses of black and grey flint. The Chipoco Formation is based on ammonites of the genus Idoceras and Glochiceras, the formation is deposited in a sedimentary, marine, shallow shelf environment and the age of the Pimienta Formation was obtained by the faunal content which includes species such as: Calpionella alpina, echinoderm remains and tintmids. According to its fauna and lithology, it is inferred that it was deposited in platform conditions with communication to the open sea and an important contribution of fine temgens, with a lithological change towards the top which is explained by a variation in the bathymetric conditions corresponding to shallow seas of low energy; the above lithological descriptions give us a general picture that the basin in the central part was invaded by marine incursions caused by the Late Jurassic transgression that continues until the end of the Early Cretaceous.

1.4 LOCAL GEOLOGY

In the study area outcrops sandstones and shales corresponding to the Lower Jurassic Sinemurian-Pliensbaquian Huayacocotla Formation, with reworked fossils, which overlies limestones and shales of the Middle Jurassic Callovian Tepexic Formation, due to the hermitage harbour saddle, evidence of a compression event. Also visible are basalt spills of the Atotonilco El Grande Formation, of Piacenzian-Gelansian age. Figure 10

Figura 10. The stratification of the weathered Huayacocotla Formation is observed.

In the study area, a strong oxidation with a branching of Stockwork quartz veins is observed with the naked eye as a consequence of the circulation of hydrothermal fluids that were conducted through a network of fractures close to the seafloor, altering the bedrock and introducing the existing mineralisation as seen in Figure 11. A large amount of disseminated pyrite is also observed, Figure 12.

Figure 11. Stockwork mineralisation can be seen.

Figure 12. Pyrite dissemination is observed.

In the stratigraphic correlation shown in Figure 1, where A) is located NE of the study area, in the direction of Xochicoatlan, where the Tepexic Formation, Tlanchinol Formation and Atotonilco El Grande Formation outcrop, with the absence of the Huayacocotla Formation in this area. B) this column lies to the SE, towards Coatitlamixtla, where the Huayacocotla Formation, Tepexic Fm. and Atotonilco El Grande Fm. outcrop. C) this column is to the SW, where the study area is located, here outcrops the Huayacocotla Formation, Tepexic Formation, and Atotonilco El Grande Formation. D) this column is to the NW, in the direction of Molango, where the Huayacocotla Formation, Tepexic Formation and Atotonilco Formation are also observed.

A Middle Jurassic transgressive cycle of the Tepexic Formation is observed in a NE-SW direction, with the smallest thickness in columns B and C, increasing in the direction of column A, against column D, where the greatest thicknesses with continental affinity are exhibited. Possibly of Precambrian origin, and continuing into the Middle Jurassic in the Cahuasasas Formation, it is worth mentioning that another N-S transgressive phase is notable, which coincides with Precambrian and Middle Jurassic normal faults located at Ixtlahuaco and
in orthogneis.

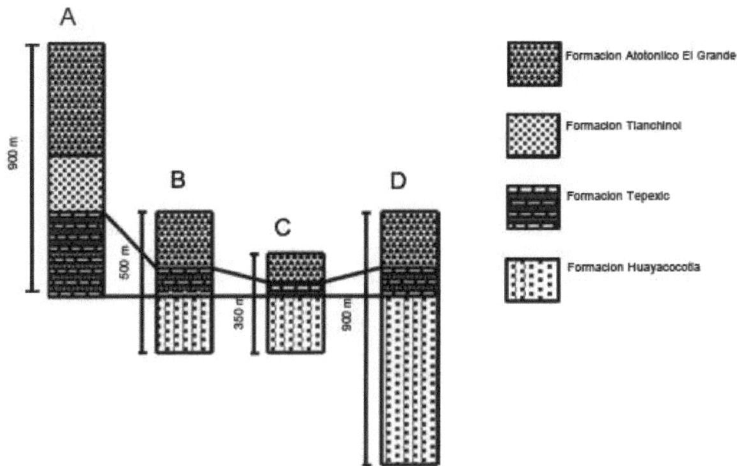

Figure 13. Stratigraphic correlation of the Xochicoatlan area, Hidalgo.

1.5 TECTONICS

Maynard and Klein [27] who build a subsidence model; Indeed, in three different areas subjected to subsidence effects Witwasterand, South Africa (Au), White Pine, Canada (Cu), and Molango, Mexico (Mn) study sedimentary basins as a tool for mineral prospecting purposes, because previously this technique has only been used to locate oil and gas, the mentioned authors demonstrated based on geochronological studies that the formations outcropping in the different mineralised districts have high efficiency for the search of inorganic minerals. In particular, in the Molango area, they describe that they observed a basin-shaped structure with fault-controlled lineaments distributed throughout the region, and calculate an extension of 1.1 for the initial rift period culminating up to 1.3, and with subsidence radii of 50 m/my with which they prove a large episode of differential subsidence that coincides at its base with the greatest tectonic activity at the base of the basin, and mention that the manganese mineralization is related to the beginning of subsidence and the onset of an incipient marine incursion of an oxidizing nature from the Gulf of Mexico combined with the presence of dissolved manganese in the reducing waters of the basin. This is important because in this work the subsidence stage that the authors refer to, accompanied by greater tectonic activity, corresponds to the Huayacocotla Formation where the mineralization of an exhalative sedimentary deposit (SEDEX) studied here is located. However, it is necessary to point out that there is ambiguity regarding the fact that there are at least two subsidence events and the Molango district is more accurate to speak of a transgressive effect than of subsidence.

Meanwhile, three main tectonic events of the Molango rift are a pre-Laramidic tectonic phase with extensional stress dominance, a Laramidic compressional phase, and finally, a post-Laramidic phase.

Thus the extension phase started during the Lower Jurassic and ended at the end of the Middle Jurassic. This led to the formation of tectonic pillars and pits, which are bounded by NNW-SSE and N-S oriented normal faults. All observed Lower Jurassic faults are bounding outcrops of the crystalline basement. During the Middle Jurassic, distinctive deformation continued, which partly led to the formation of new pits and tectonic pillars, evidenced by the abrupt lateral termination of the red beds of the Formation, indicating that they were deposited in a pit system.

The Jurassic and Cretaceous sedimentary rocks were folded between the Upper Cretaceous (Laramide Orogeny), forming a complex of thrust folds locally called the Huayacocotla Anticline, whose complexity is caused by the pre-existing structural geometry of the basement, as well as by the horst and Grabens sediments in this region. The basal detachment of the Laramide structures is found in the Huayacocotla Formation, which rests on the Precambrian metamorphic basement. Stripping of folds and flats. Another Laramide phenomenon is the reactivation of the Jurassic normal faults as reverse faults, which delimit the Jurassic Huiznopala horst (upper basement) on its eastern side. This inversion is associated with a local maximum relative shortening. The evidence of such shortening is manifested to the east of the Huiznopala basement high, where the sedimentary cover is intensely folded, the cover above the basement high being less folded. Between the Santiago and Chipoco formations is a tectonic rift. It originated during the Laramide orogeny. The abrupt manganese partitioning at the base of the Chipoco Formation, the total absence of manganese at the top of the Santiago Formation and the abrupt lithological change between these two formations, suggest an abrupt change in the physico-chemical conditions of the trench where the Santiago Formation was being deposited, which is thought to be due to the communication between the Huayacocotla Trench and the Gulf of Mexico, which also allowed communication between the Pacific and Atlantic oceans.

The last tectonic event is due to post-Pliocene extension, evidenced by NW-SE trending normal faults. They are delimiting the Molango graben. The graben is more than 10 km long and has a 200 m drop.

The Cretaceous Jurassic sedimentary rocks were folded during the Upper Cretaceous to Upper Eocene (Laramide Orogeny), and formed a complex set of folds and thrusts (Huayacocotla Anticline), in this area it is possible to observe the reactivation as reverse faults of the Jurassic normal faults, which allow inferring a Jurassic horst in the vicinity, (basement high); this inversion could be due to a local maximum relative shortening. Evidence is inferred to the east of the basement high, where the sedimentary cover is intensely folded, whereas the sedimentary cover of the basement high is less folded. The possible copper-silver deposit is located exactly in the zone of highest structural complexity. In the vicinity of the Naopa thrust, and with intraformational thrusts of the Cahuasas Formation.

Outreach Events

During the Lower and Middle Jurassic that led to the formation of tectonic pits and pillars bounded by normal faults. In the Middle Jurassic the formation of tectonic pits and pillars continues and is evidenced by the abrupt termination of the red beds of the Cahuasasas Formation. Jurassic and Cretaceous sedimentary rocks were folded during the Late Cretaceous-Late Eocene, in what is known as the Laramide Orogeny, forming a complex set of folds and thrusts called the Huayacocotla Anticline. Finally, there was a post-Pliocene extensional tectonic event, evidenced by normal faults that are delimiting a graben as the Molango graben.

1.6 Rift Molango

It is a topographic depression formed by subsidence effects caused by relatively parallel normal faults, which are associated with volcanic and seismic activity.

Geosynclines develop only at the margins as a result of the separation along active rift arms. On the other hand, an Aulacogene is a tectonic structure that is initiated by an extension phase in a certain region, when this phase ceases and does not present a well-formed tectonic aspect, it is called an Aulacogene. Between both structures, rift and Aulacogene, there is a difference that allows to recognize them, the rift continues with an extension phase after its origin, the Aulacogene starts as a narrow graben, limited by faults, later it widens, the direction of the stresses produces dipping faults and it is also broken by faults.

In order to recognise the presence of the main tectonic structure in Molango, Hidalgo, the following criteria were also taken into account:

1)   Basalt flows and bimodal ^gneiss rocks

Molango has a sequence of a bimiodal nature (Robin 1979) that includes the Tlanchinol and Atotonilco el Grande Formations, both of Cenozoic age.

2)   Fault-controlled guidelines

Suter (1990) and Carrillo-Martinez (1991) described the tectonics of a portion of the Sierra Madre Oriental and part of the Tampico-Misantla Basin; they also describe northwest-southeast fault-controlled alignments; Ochoa Camarillo (1996, 1997) describes northwest-southeast and north-south fault-controlled alignments in the central part of the Huayacocotla Anticlinorium.

3)   the pillars and pits create a palaeotopography of basin and mountain ranges the topography of basins and mountain ranges is manifested in this area as the result of the pre-existing structural geometry of pits and pillars, with some faults delimiting pillars and tectonic pits, another vestige is provided by the pilcuautla shaft that cuts the basement.

4)   Lateral facies changes

In the Molango rift, there are lateral facies changes over short distances as in the Middle Jurassic with two lithologies that form the Cahuasasas Formation, as well as the Upper Jurassic Chipoco Formation that changes laterally to the Taman Formation.

5)   Change of sedimentation

Rapid rate, rapid sedimentation rate is observed in Palaeozoic formations (Otlamalacatla and Tuzancoa) east of Molango and in the Lower Jurassic Huayacocotla Formation. More noticeably, rapid sedimentation is observed in the Middle Jurassic red-bed sequence of the Cahuasasas Formation. Slow rate, representing the beginning of a basin sedimentary cycle in a shallow environment and continuing with the clayey Santiago Formation and the Upper Jurassic Chipoco Formation.

In limited areas of the Molango region, the Chipoco Formation overlies the Pepienta Formation, which represents an advanced stage of basin sedimentation or subsidence stage, which continues with the lower Tamaulipas Formation. These slow sedimentation rate formations have been considered to varying degrees as hydrocarbon bearing rocks.

6) Magnetic anomaHas Magnetic chart F14D51 contains evidence of magnetic anomaHas.

CHAPTER 2

METHODOLOGY

2.1 Working method

The present work was carried out on the basis of an extensive compilation of bibliographic data from the area of Molango, Hidalgo, which served to know the stratigraphic and tectonic characteristics of the area. We were able to take several reference sites as strategic points for field data collection, which later allowed us to correlate and compare tectonic and stratigraphic data and to know the paleoenvironmental characteristics of the area.

Knowing the paleogeographic characteristics allows us to know the pattern of the transgressions that affected the territory of Hidalgo during the Liasic. This information was very useful for the planning of field trips for the reconnaissance of the area, data collection, sampling and mining prospecting. On the other hand, based on this information, we were able to define points favourable to mineral settlement, giving as a consequence the location of mineral deposits of economic interest.

A series of field trips were made for the reconnaissance of the area, visiting the area described by the authors consulted and locating the different lithologies described in the area as well as the type locality of the Huayacocotla Formation, an area of mineral interest. Likewise, in some of these sections stratigraphic surveys were carried out for descriptive purposes for the following work. It is worth mentioning that a point where the Huayacocotla Formation outcrops is proposed, an area where a specific stratigraphic survey was carried out and a series of samplings including drilling in the area.

By counting the stratigraphic and structural data we were able to determine the tectonic affinity of the region and infer which environments were predominant during that time and which mega structure corresponds to this type of lithological and dynamic variables. The above data allow us to locate exploration targets.

2.2 Location and access routes

The municipality of Molango and its surroundings are part of the study area, it is located in the northern part of the State of Hidalgo at 20°40' - 21°05' N and 98°33' - 98°50' W at an altitude of 1620 masl. It is located in the Huayacocotla Anticlinorium, which is part of the belt of folds and ridges of the Sierra Madre Oriental.

Federal Highway 105 Mexico-Huejutla de Reyes is the most important access road to the study area, which is located in the vicinity of the municipality of Molango, with emphasis on the town of Xochicoatlan, located to the northeast of the municipality.

Figure 14. Location and access roads to the study area.

2.3Fieldwork

2.3.1 Huayacocotla Formation

The Huayacocotla Formation is located at coordinates N 20°44.644'y W98°42.566' at an altitude of 1524 m above sea level, this outcrop is approximately 15 m high and is composed of red lutites in the open and black in the fresh air arranged in slabs, the entire outcrop is highly altered, has areas of milonitizacion by the action of tectonic and weathering by adverse agents as seen in the figure

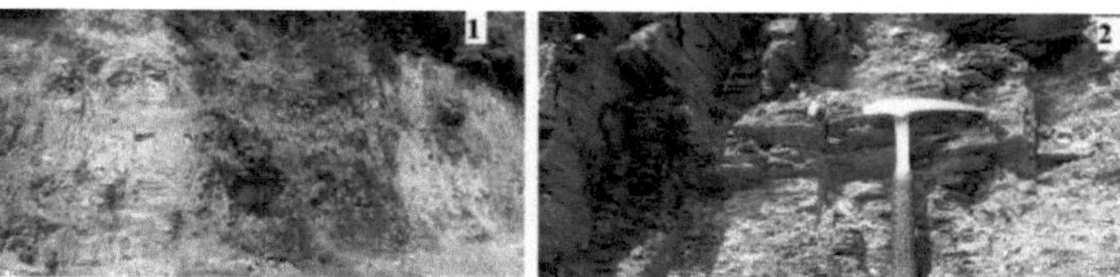

Figure 15, Huayacocotla Formation outcrop shale member, and in 2 Huayacocotla Formation outcrop with shales in slabs, hammer as 30 cm scale.
It is worth mentioning that this outcrop is not of economic interest as this is not the member where the SEDEX-type mineralisation of the Huayacocotla Formation.

Figura 16.   Lutites in slabs of the Huayacocotla Formation. (1) Black frescoed slabs, five Mexican peso coin, scale 3 cm approximately, (2) Plans of the lutites in pencil as scale of 10 cm approximately.
This outcrop is located at coordinates N20°45.983' and W 98°41.863'.
With a latitude of 1767 masl, this portion of the Huayacocotla Formation is composed of centimetric to metric strata of fresh grey sandstones and weathered ochre and dark grey shales in strata.
The outcrop is clearly affected by the tectonics of the area, showing abrupt N-S fault folding.

Figura 17.   (I) Outcrop of the Huayacocotla Formation, (2) sandstone strata with shales, (3) hand sample of a shale.

The Huayacocotla Formation in this area is represented by the third member described by Imlay [34] and according to this study its environmental conditions at the time and general tectonics of for the emplacement of minerals characteristic of a SEDEX type deposit.

Morphological evidence such as silicified stockwork material and a contact with fragments and vertical pseudo-stratification showing traces of black smokers, fossils found in the area, as well as mineralogical evidence such as the occurrence of jarosite and disseminated minerals such as pyrite were collected in the field, later verified by geochemistry of the collected samples.

Figura 18.	(I) stockwork type mineralisation, (2) jarosite, (3) disseminated pyrite, (4) pyrite, (5) pyrite, (6) pyrite, (7) pyrite, (8) pyrite, (9) pyrite.

In this area, a drill hole was drilled to obtain an auger core as shown in figure 19. In fact, to choose the location of the drill hole we first considered a place where the stratigraphic characteristics correspond to rocks of Lower Jurassic age, we also considered the criterion of locating a fault zone related to tectonic pillars possibly pre-Cambrian that formed the pre-existing local topography, in addition, we considered this area for drilling, because of the observed mineralization of exalites, in addition to a typical stockwork structure.

Figura 19.   sampling diamond drilling.

It is worth mentioning that the selected outcrop is about 10 to 100 metres thick, it is made up of black shales intercalated with sandstones. The black shales contain intervals of disseminated pyrite, the pyrite horizons have a sedimentary structure, some strata are altered, and oxidation zones are observed with iron and manganese minerals, and high silicification with pyrite, An important factor is the occurrence of normal faults between the Precambrian gneiss or Paleozoic volcaniclastic and volcaniclastic rocks during the sedimentation of the Huayacocotla Formation, with zones of strong silicification with pyrite.

The Tepexic Formation is located within the coordinates N 20° 444.977' W 98° 42.957' at 1530 masl and is composed of grey limestones of metric strata. This formation represents the beginning of carbonate deposits in the area, marking the change of environment in the area to an oxidizing environment, for these reasons it can be said that the Tepexic Formation represents the base of a marine sequence.

Figura 20.   Outcrop of the Tepexic Formation, scale person approx. 1.60m scale.

2.3.3 Training Santiago

The Santiago Formation is located at coordinates N 20°46.697' and W 98°43.563' at

Figure 21, Santiago Formation outcrop, scale person approximately 1.75m, Santiago Formation shales hammered to approximately 30cm.1767 masl,
is composed of shales with a high fossiliferous content.

2.3.4 Pepper Formation

The Pimienta Formation is located within the coordinates N 20°42.983' and W 98°42.190' at 1731msnm. It is composed of a lithological sequence of centimetric strata of weathered grey and black limestones interbedded with carbonaceous shale, calcite veins and flint nodules. It is affected by a series of faults and folds that cross it due to stresses caused by the opening of the Gulf of Mexico and the compression of the Laramide orogeny. This formation contains

Figure 22, Panoramic outcrop photograph of the Pimienta Formation, (II) Close-up photograph of black limestones and carbonaceous shales of the formation, hammer approximately 30cm and (III) calcite vein, Mexican fifty centavos coin.

characteristic chevron-type trailing folds
A series of chevron folds and a sequence of NW-aligned faults can be observed along the formation. Two faults in the outcrop were measured and the following data were obtained: fault 1 has a strike of 20°NW and a dip of 18° SE while fault 2 has a strike of 60°NW and a dip of 20° SE.

Figure 23, Panoramic photograph of the chevron folds of the Pimienta Formation, (II) Close-up photograph of the chevron folds.

2.3.5 Chipoco Formation

The Chipoco Formation is located at coordinates N 20°43.539' and W 98°42.982'. It is made up of strata ranging from centimetric to decimetric manganese spherical limestones of black and reddish opaque colour when weathered and black to deep purple when fresh. In this formation manganese minerals such as rhodochrosite, kutnaurite and pyrolusite are found. Some authors consider it to be

Figure 24, (I) outcrop of the Chipoco Formation, (II) manganeseiferous limestone, approximately 10cm pencil.

as the formation that marks the opening of the Gulf of Mexico.

2.3.6 Tlanchinol Formation

The Tlanchinol Formation is composed of massive basalts of reddish colour in weathered samples and black when fresh. They are of great interest for this study because bimodal ^gneous rocks represent one of the parameters of Jowet (1986) that are useful for the identification of an ancient rift.

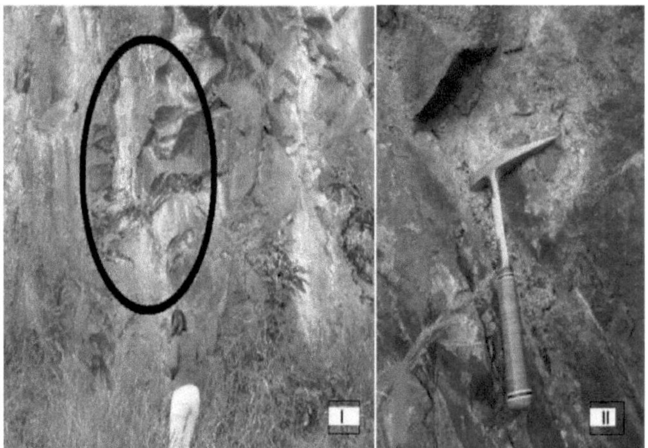

Figure 25, (I) Tlanchinol Formation outcrop, (II) massive basalts, hammer scale about 30cm.

2.3.7 Atotonilco el grande Formation

The first data collection was carried out at coordinates N 20° 20'11.2" and W 98°42'24.9" with an altitude of 2006 masl in the area of San Jose, Hidalgo, in an outcrop of the Atotonilco El

Grande Formation; the outcrop presents a large amount of caliche in the fractures, this represents evidence of recent tectonic movements that may not be more than 10,000 years old and calcium sulphate minerals and possibly jarosite. Some hand samples contain dark minerals, which is important in the case of sulphates because they indicate an extremely acidic pH, approximately pH 1.2 during their formation. It is important to mention the height of sea level during the formation of these limestones, which is 2006 m, as it is inferred that this formation outcrops at 756 m below according to data collected in the field. The following photograffa, which we call A, shows in the outcrop where we observe fractures that have a NW-SE course and the preferential lineation of the structures N-S, in the outcrop presents propylitic alteration in the fracture and diagenetic alterations of chlorite, in the mentioned outcrop different levels of deposits are observed: I) at the base there are interdigitated limestone deposits with calcite stylolites parallel to the stratification, at the same level there are poorly sorted clasts and slope deposits. II) shows more hydrothermal alteration in which chlorite distribution begins. III) the distribution of clasts is more homogeneous and has a better distribution of chlorite. IV) In this zone the temperature at the time of formation was higher, so propylfic alteration is abundant.

Photographs were taken of three sections of the outcrop, in which the alterations characteristic of the different stages were observed. Photograph B shows poorly sorted clasts of approximately 2 cm, found in level IV of the outcrop. Photo C is a semi-detail image showing diagenetic alteration and alteration in the clays by hydrothermalism, leading to the production of chlorite, which is best observed in level III. Finally, in image D, limestone is observed with stylolites parallel to the stratification present in level I.

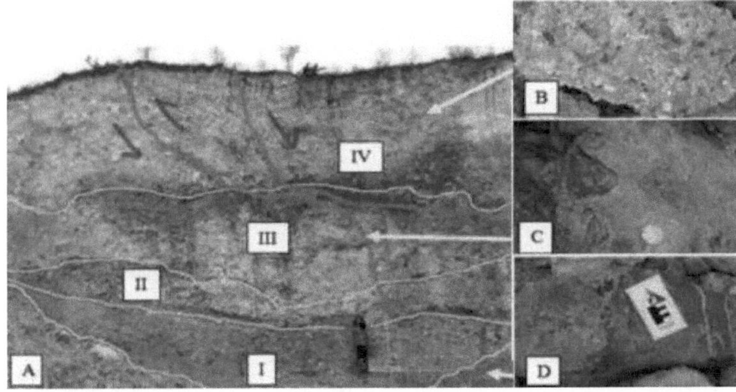

Figure 26, (A) Outcrop of the Atotonilco El Grande Formation, (B) Semi-detail image of part IV of the outcrop, (C) Semi-detail image representative of hydrothermalism and diagenetic alterations, (D) Semi-detail image of limestone with stylolites, (E) Semi-detail image of limestone with stylolites, (F) Semi-detail image of limestone with stylolites.

2.3.8 Cerro colorado

In the locality of Cerro Colorado, Hidalgo, at coordinates N 20° 24' 29.4 and W 98°41 '43.4" with an altitude of 1872 masl. Several features were observed and will be described below. In Figure 27, image E shows an outcrop with ejecta foci (I), pseudostratified rhyolites (II), diaclases and fractures with NW-SE direction (III), lineaments with N-S direction. Meanwhile, the escarpment zone has an inclination of 82° and it is worth mentioning that there are slope deposits of approximately 1.50 m. The following is a more detailed description of each of the aforementioned elements. Image F corresponds to an enclosure of the ejection focus, using the pike as a scale. A mixture of volutes, volcanic material and rhyolite is observed, the fluid material forms a halo with a hemicircumference of 7 m and a halo thickness of 10 cm of rhyolite. It is worth mentioning that in this outcrop a differential thermal gradient indicator is observed and the thickness of the halo is approximately 10 cm.

Figure 27, (E) General outcrop image of the Cerro Colorado locality, (F) outcrop structure, (G) rhyolites at the periphery of the outcrop hammer as a scale of approximately 30cm, pencil as a scale of approximately 10cm.

2.3.9 Leon Pass

The locality of Paso de Leon is located at coordinates N 20° 25'56.4" and W 98°41'06.2" at an altitude of 1737 masl. In this outcrop, figure 28, hand samples were obtained where kinematic slip indicators (striations) were observed. The outcrop is divided into three levels; level I is the cornice front or escarpment, level II is made up of slope deposits and level III is made up of a distribution of volcanic rock, Tezontle with Irinzite minerals.

Figure 28 Paso de Leon igneous outcrop.

2.3.10 Acalome.

In this locality the abrupt tectonic contact is observed, in it the direction of the compressional stresses are indicated with dates that make it look like a saddle where the rocks of Upper Cretaceous age underlie the Tertiary. A tectonic calcite breccia showing calcite intergrowth is also observed, the contact is mylonitized and contains fine fragments of material resulting from low-grade metamorphism.

Figure 29. (J) Tectonic contact image, in I Upper Cretaceous rocks, II Tertiary rocks, (K) brecciatectonic with calcite clasts.

2.3.11 San Agustin Metzquititlan.

The Atotonilco el Grande Formation is composed of columnar basalts of approximately 20 m in height.

Figure 30. Columnar basalts of the Atotonilco El Grande Formation.

2.3.12 The Bank

In the locality the bench is an outcrop of Met a s s o mat i c , which at the base is composed of volcanic glass perlite and is capped by rhyolitic spills.

Figure 31. Metasomatic outcrop in the Atotonilco el Grande Formation.

SAMPLING

Sampling was carried out with the aid of a Winkie Portable drill rig, which is a 2-cycle gas powered, air-cooled drill rig, drilled at the coordinates, with an inclination of 45°, at coordinates 20° 46' 01" N and 98° 41' 86". A core of 9 metres in length and with a diameter of 1.23 inches was extracted in the study area, which was previously programmed after having carried out the recognition of the geology of the area, as well as the structures present, in addition with the help of the mineralogy it was possible to infer the site where it would be of interest to drill to know a little more about the deposit, inferring on the basis of the reconstruction paleogeograffa the presence of base metals in the deposit.

Figure 32, Drilling machine in the study area.

The drill core was sampled every 0.25 m and 36 samples were obtained, of which 9 samples were characterised by the different geochemical methods.

Table 1, classification and depth of samples

Number of samples	Name of samples	Depth (cm)
1	XOCH-100	100 cm
2	XOCH-200	200 cm
3	XOCH-300	300 cm
4	XOCH-400	400 cm
5	XOCH-500	500 cm
6	XOCH-600	600 cm
7	XOCH-700	700 cm
8	XOCH-800	800 cm
9	XOCH-900	900 cm

It was also possible to identify some minerals in auger samples, such as quartz, monazite, pyrite, chalcopyrite, and muscovite Figure 33.

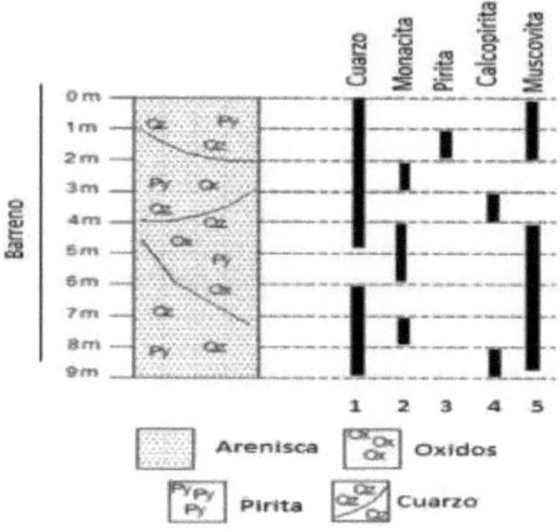

Figure 33, relative distribution of minerals in the borehole.

2.4 X-RAY DIFFRACTION ANALYSIS (DRX)

The X-ray diffraction technique is very useful for identifying mineral phases; a wide variety of data can be obtained, such as: the crystalline system of the different minerals, their approximate composition, crystal size, crystallinity, preferential alignment, among others. Due to its versatility and the large amount of data it provides, it is a very popular technique.

The source or tube used generates cobalt radiation, and has a curved detector architecture using an inert gas mixture such as ethane-argon.

The operating power of this equipment, (Figure 34), is usually 20 mA - 30 Kv. On the other hand, the sample is prepared to a fine size of 35np or 200 mesh of Taylor's series. Using an agate pestle and mortar to pulverise the sample, which weighs approximately 0.5 grams each.

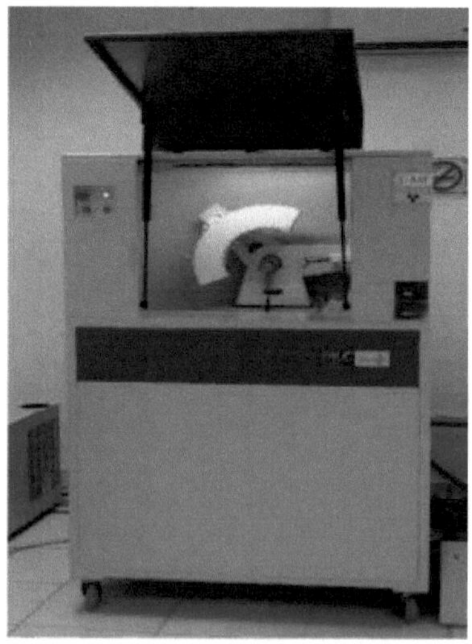

Figure 34, Inel Diffractometer, Equinox 2000 model.

2.4.4 Sample Preparation

X-ray diffraction (XRD) analyses on core samples were prepared by pulverising the sample in an agate mortar to a particle size of approximately -10 pm, then compacting it in an aluminium sample holder approximately 1 cm in diameter, supported by a 2.54 cm long Al piston.

The sample as^ mounted; was placed in an Inel diffractometer, model Equinox 2000 equipped with a curved detector operable with a 99.9 % high purity ethane-argon gas mixture. This type of analysis was carried out on the mineralised zone, and the data obtained from the diffractogram was compared and evaluated with Match 1.0 diffraction software.

2.5 SCANNING ELECTRON MICROSCOPY AND ENERGY DISPERSIVE SPECTROMETRY ANALYSIS (MEB-EDS)

Scanning electron microscopy SEM (figure 35) provides the texture and morphology of the mineral sample. In addition, if EDS is available, it is possible to obtain a rough elemental analysis at a precise point. If you need to observe the texture of the mineral, existing micro fossils, their texture, or observe a halo of decomposition, then you should cut a small piece of sample, mount it in resin, and grind and polish it to a mirror finish. If you want to observe the sample without bisecting any particular structure, you can mount the powder on a special SEM tape. In both cases, specimen or tape, you should coat with graphite if you want to determine metals including gold, and you should coat elements with low conductivity with gold. There is no need to coat if the SEM is of a low vacuum.

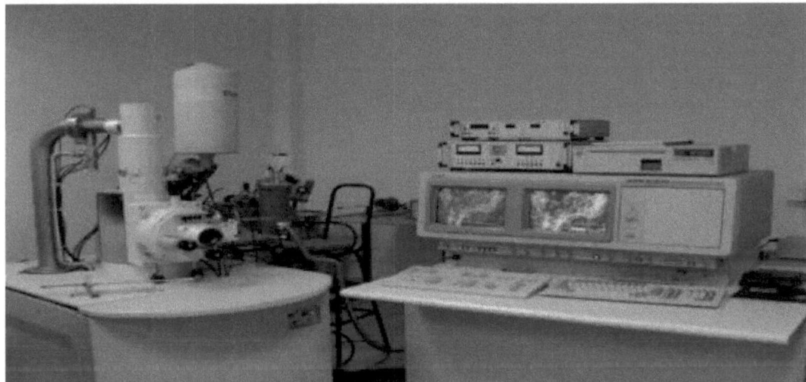

Figure 35. JEOL JSM-6300 microscope and equipped with an EDS detector.

3.5.1. Preparation of the sample

The samples analysed by scanning electron microscopy were used to study textures and qualitative distribution of the elements in the core samples.

Therefore, the samples to determine the composition by EDS were finely pulverised mineral quarticulated (at +200 mesh or more than 35 mp ground with agate mortar, mounted on adhesive aluminium foil and uncoated; subsequently the collected core samples were cut and coated with gold to be analysed on a JEOL JSM-6300 microscope and equipped with an EDS detector.

43

2.6 INDUCTION COUPLED PLASMA ANALYSIS (ICP)

Inductively Coupled Plasma-Mass Spectroscopy (ICP-MS) is a high resolution method that allows the detection of element concentrations in the order of parts per million and from there they are transformed into other units such as ppm, gr/tonne, or are transformed to oxides. It is an instrumental analysis technique that shows the chemical analysis per element in ppb.

Wet chemistry is, of course, a traditional and very accurate chemical analysis technique. Using reagents commonly used in chemistry, various volumetric techniques are used to calculate indirectly the amount of a particular chemical element and to provide with high accuracy the percentage of the chemical element.

In many chemical analyses, 1 gram is sufficient to perform the analysis in triplicate. By simple rules of three, the percentage of the element is determined as an element or as a mineral according to the volumetric consumption and requires a variety of high purity chemical reagents, some acids and bases, as well as laboratory glassware.

2.6.1. Sample Preparation

From one gram of sample, dilutions are required according to the specific technique, standards or solutions with a specific element content are prepared for the calibration curve; usually, ultra-high purity argon gas is used to generate the plasma.

CHAPTER 3
RESULTS

For the prospection of the SEDEX deposit of this work, a series of walks were carried out where samples and stratigraphic and tectonic data were taken. In order to know the tectonic affinity, we analysed the marine transgressions, the stratigraphic setting and the tectonics of the area, which provided us with the evidence according to the criteria of Jowett [40] to delimit that the study area is located in a basin of ancient rift type extension.

Stratigraphic analysis showed that SEDEX-type mineral mobilization lay during the Lower Jurassic and was emplaced in the Huayacocotla Formation during marine transgressions during the ritfting stage.

In order to recognise the mineralogical characteristics, a series of geochemical analyses were carried out, correlating the results with the zones previously mapped by various authors, thus corroborating their environment and formation conditions.

The XRD analysis of each of the formations allows us to know the mineral phases that predominate in each of the formations studied here, which allows us to know the environment and formation characteristics of each one of them.

Once the point for the formation of a SEDEX deposit was delimited, and a drill core was extracted where ICP analysis was carried out, it provided us with quantitative elemental results which give us characteristic parameters of a SEDEX deposit rich in elements such as Pb, Zn, Cu, Ag, Au, Bi, Ba. On the other hand, in

SEM and EDS analyses show pyrites are amorphous botroidal-type Hpica pyrites from exhalative fumes to which the change in the deep basin environment could be attributed.

3.1 TECTONIC AFFINITY

In particular, a rift is a large fissure in the earth's crust, which starts as a small fissure derived from stresses generated in the tectonic plates, and as they interact with each other mainly in subduction zones, they cause a geodynamic movement within the crust. Therefore, over millions of years in the geological history of a rift, several types of mineralisation can be associated during or after the rift sequence, and due to the phenomena associated with the stresses generated during the formation of the rift as shown in the diagram in figure 35. In the study area, thicknesses were measured from 350 m to 900 m of the Huayacocotla Formation and thickness of The producer horizon emerged at 100 m, and the upper unit of shale with organic matter outcrops at 250 m of sedimentary rocks and increases in some areas with outcrops of volcanic rocks at 350 m, the intermediate portion is greater than 250 m in thickness, while the basal portion was observed at about 450 m. Figure 36, in [18], also shows the lithological correlation of the Lower Jurassic transgression.

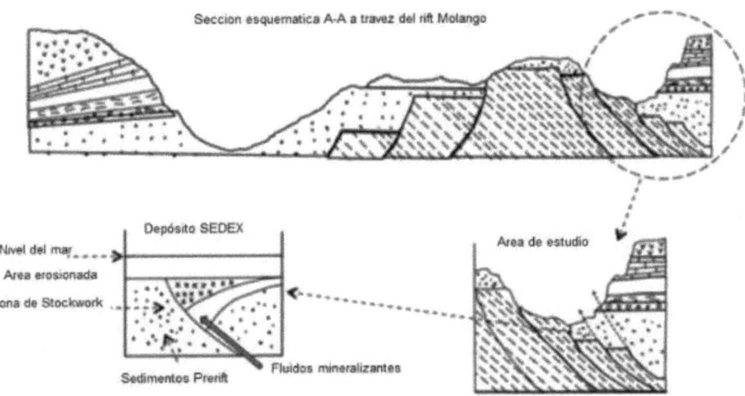

Figure 36, Schematic section through the Molango rift.

Meanwhile, the possible mechanism for the present-day configuration of the Molango rift is due to the pre-existing structural geometry perhaps dating back to the Proterozoic with the formation of tectonic pillars triggered by compressional stresses causing their emersion.

On the other hand, during the Jurassic, extensional stresses predominate and continue this trend towards the end of the Upper Jurassic; but the stresses changed direction to compression in the Cretaceous during the Laramide Orogeny, to finally reactivate extension in the Pliocene. Therefore, the lithology changed in nature, so that the Lower Jurassic in the area is predominantly marine, the Middle Jurassic continental, the Upper Jurassic, Cretaceous marine, and finally the Pliocene continental.

Figure 36, schematic depositional arrangement of the formations outcropping in the rift core.

On the other hand, it is worth mentioning that the basin studied in this study was denominated as a riff type basin according to Jowett's criteria [6], which were fulfilled in the following way:

Basalt flows and bimodal ^gneiss rocks

Both the Tlanchinol Formation and the Atotonilco El Grande Formation consist of bimodal rocks. The former is made up of basaltic flows and effusions intercalated with felsic sequences (rhyolites) and other Robm pyroclastics [38] while the Atotonilco el Grande Formation is made up of basalts intercalated with tuffs of rioKtic, dacitic and andesritic composition.

Long fault-controlled lines

The most outstanding geological structures in the area are delimited by faults running NE-SE, such is the example of the Naopa saddle and Molango graben, Ochoa Camarillo [34] mentioned fault-controlled lineaments to the fold system that he called the Anticlinorio de Huayacocotla when describing the structural behaviour of the Sierra Madre Oriental he mentioned lineaments that behave in the same way.

Topography of pits and pillars generating basins and ridges

The study area was affected by extension and compression events that generated the present topography where there are basins delimited by mountain ranges whose structures correspond to grabens and horst.

This can be checked through the evaluation of the density of the structures which gives an indication of the cortical strength or weakness of the study area. On the other hand, the change in slope is indicative of the tectonic nature of a structure and shows possible relationships with faults or fractures. Meanwhile, in the study area, the analysis of the main structures in Figure 37, their tectonic analysis, and the erosion resistance to establish the geochemical dispersions were carried out. It can be seen that most of the stresses are positive, with a greater occurrence of horst-type structures, which form pillars.

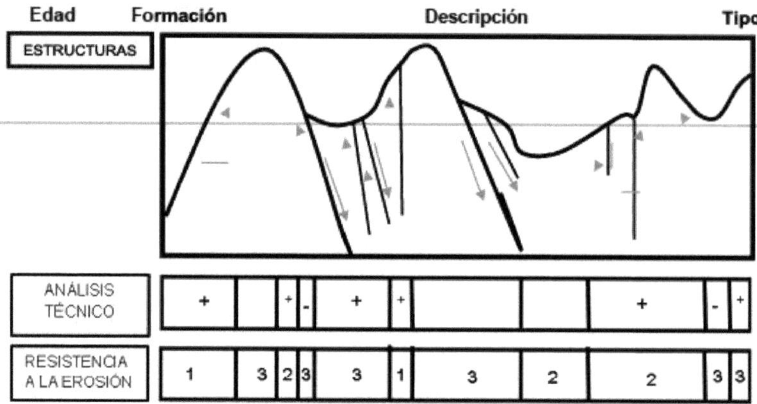

Figure 37. Erosion resistance analysis.

Likewise, according to the nature and degree of erosion for each formation, it is important to evaluate the relationship with tectonic structures because an incompetent rock erodes faster and the scattering halo can be larger. The formations were classified in table 2 into three types of erosion resistance, type 1, high resistance, type 2, medium resistance, and type 3, low resistance.

Table 2, Erosion resistance

Value	Erosion resistance
1	BAJA
2	MODERATE
3	ALTA

Also, in table 3, the relation of lithology to erosion shows that harder rocks are less susceptible to erosion, thus basalt is number one, or low erosion, while high erosion rocks are exhibited by the clayey Santiago Formation.

Table 3. Lithology/erosion ratio

Miocene Top	Tlanchinol	Bimodal lava spills	1
Jurassic Top	Pepper	Light-coloured mudstones, clean or with little clay, planktonic microforaminifera and flint lenses.	3
Jurassic Top	Chipoco	Manganese^bearing limestones with intercalations of calcareous shales.	3
Jurassic Top	Santiago	grey, weathering to brownish-brown or reddish-brown calcareous shales, with intercalated calcareous nodules.	3
Jurassic Medium	Tepexic	Grey to dark grey, coarse-grained, impure limestones with many quartz grains.	2
Jurassic	Cahuasas	Sediments of continental origin consisting of	2

Medium		sandstones, conglomerates and red beds.	
Jurassic Inferior	Huayacocotla	banded, shaley and fractured black shales also containing some thick sandstone strata	2

To calculate the erosion factor as shown in table 4, corresponding to the different lithologies, the exposed surface area according to its type was calculated between the estimated channel length. Calculated as follows per Knea of outcrop (field).

Table 4. Calculation of erosion factor A

neaA	Surface area (Has)	From right to left $^{^2}$ $^{-1}$	Factor
1	100 /	[3.3 X 500]= 606.061 m(1 ha/10,000 m) =	0.06
2	100 /	[0.9+0.6+0.5+0.3+0.5+0.4+1+0.5+0.5 * 500]	0.04
3	100 /	[2.5+0.8+1.4*500]	0.04
4	100 /	[1+0.8+0.5+*500]	0.09
5	100 /	[2.8+2.4+1*500]	0.03
6	100 /	[0.8+1.4+0.5+0.2+0.8+1.4*500]	0.4
7	100 /	[2.2+1.5*500]	0.5
8	100 /	[0.5+1.2*500]	0.12
9	100 /	[0.9+1.5+1.3*500]	0.05
10	100 /	[1.6+1.9+0.6+1+0.9+0.1*500]	0.05
11	100 /	[1+0.4+1 +1.7+0.2+0.5+0.6+1.3*500	0.03
12	100 /	[1 + 1.2+0.3+0.5*500	0.07

1=100has

Table 5. Calculation of erosion factor B

Lmea (Field) B	Surface area (Has)	From left to right	Factor
13	100 /	[0.5+1.2+0.3+0.7 * 500]= (1 ha/10,000 m) = (1 ha/10,000 m) =	0.07
14	100 /	[1+0.3+1.7+0.5* 500]	0.06
15	100 /	[0.4+0.2+1.4+0.7+0.5+1.2+1*500]	0.04
16	100 /	[0.8+0.3+0.6+0.5+1.5*500]	0.05
17	100 /	[0.9+1.0+0.8*500]	0.07
18	100 /	[0.7+0.8+0.3*500]	0.11
19	100 /	[0.5+1.3+0.2+1.6+0.8*500]	0.05
20	100 /	[0.5+1.5+2.2+0.5+0.8*500]	0.04
21	100 /	[1.2+0.8+0.6+1.5*500]	0.05
22	100 /	[0.7+1.1 + 1+0.4*500]	0.06
23	100 /	[2+1.3+1.5+0.7+0.8+0.7*500]	0.03
24	100 /	[1+0.5+1 + 1.2+1.2+1.8*500	0.03

Table 6. Calculation of erosion factor C

Lmea (Field) C	Surface area (Has)	From left to right i	Factor
25	100 /	[2.2+1.8+1.6+1.6+0.3 * 500]= (1 ha/10,000 m) = (2.2+1.8+1.6+0.3 * 500]= (1 ha/10,000 m) =	0.03
26	100 /	[1.2* 500]	0.17

27	100 /	[2.0+1.0+0.7+0.7*500]	0.05
28	100 /	[1.0+0.4*500]	0.14
29	100 /	[1.5+0.7*500]	0.09
30	100 /	[0.5+1.5*500]	0.10
31	100 /	[1.0*500]	0.20
32	100 /	[2.2+1.7*500]	0.05
33	100 /	[2*500]	0.10
34	100 /	[1.5+0.7*500]	0.09
35	100 /	[1.0+2.2+1.5+1.0*500]	0.04
36	100 /	[1.9+1.5+1.7+1.3+1.2+0.7+0.3+1.2*500	0.02

Table 7. Calculation of erosion factor D

Lmea (Faithful) D	Surface area (Has)	From left to right	Factor
37	100 /	[0.8+1.2 * 500]= (1 ha/10,000 m) = (1 ha/10,000 m) =	0.10
38	100 /	[0.7+0.8* 500]	0.13
39	100 /	[0.9+0.2*500]	0.18
40	100 /	[1.0+0.6*500]	0.13
41	100 /	[1.2+1.2+1.4*500]	0.05
42	100 /	[1.8+0.3+0.5+0.8*500]	0.06
43	100 /	[1.5+1.2+1.4+4.1*500]	0.05
44	100 /	[1.2+0.6+1.1+0.2*500]	0.07
45	100 /	[1.5+0.7*500]	0.09
46	100 /	[0.3+0.6+1.4*500]	0.09
47	100 /	[1.2+1.0+0.7*500]	0.07
48	100 /	[0.5+1.4+1.8+2.0+0.2*500	0.03

Table 8. Calculation of erosion factor E

Line (Field) E	Surface area (Has)	From left to right	Factor
49	100 /	[0.7+0.5+1.0+1.0+1.2 * 500]=(1 ha/10,000 m) =	0.06
50	100 /	[1.4+0.2+1.3* 500]	0.07
51	100 /	[2.0+1.0+0.3+1.5+0.6*500]	0.04
52	100 /	[1.2+1.2+0.7+0.5*500]	0.06
53	100 /	[0.9+0.8+1.4+0.5+0.6*500]	0.05
54	100 /	[1.0+1.4+0.7+0.8+1.0+1.0+0.5*500]	0.03
55	100 /	[1.3+0.7+0.7+1.4+1.9+1.0+0.5*500]	0.03
56	100 /	[0.5+1.0+0.6+0.6+1.5*500]	0.05
57	100 /	[1.6+1.9+1*500]	0.04
58	100 /	[1.0+1.4+0.6+0.6+1.5*500]	0.04
59	100 /	[0.3+0.7+0.5+0.8+0.7*500]	0.07
60	100 /	[0.2+1.4+0.8*500	0.08

From the calculation of the factor for each area, the zones with the erosive gradients are shown in the table. The sedimentation rate is higher when:

There are decreasing values up to the minimum, 0.02, furthermore, 0.03, 0.04, 0.05, 0.06, 0.07, 0.08, 0.09, 0.10. The least eroded are: 0.11, 0.12, 0.13, 0.14, 0.15, 0.16, 0.17, 0.18, 0.19, and 0.20.

Table 9. Shows the erosion zones, note the coincidence with horst and tectonic pillars.

A	0.07	0.03	0.03	0.05	0.12	0.05	0.04	0.03	0.09	0.04	0.04	0.06
B	0.07	0.06	0.04	0.05	0.07	0.11	0.05	0.04	0.05	0.06	0.03	0.03
C	0.03	0.17	0.50	0.14	0.09	0.10	0.20	0.05	0.10	0.09	0.03	0.02
D	0.10	0.13	0.18	0.13	0.05	0.06	0.05	0.07	0.09	0.09	0.07	0.03
E	0.06	0.07	0.09	0.06	0.05	0.03	0.03	0.05	0.04	0.04	0.07	0.08

4. Lateral facies changes.

Chipoco-Taman formations

Cahuasas change of litholog^a red-bedded conglomerates at Ixtlahuaco and at Tepehuacan red-bedded sands and oxidising conditions.

Laps of 20-25 m.a of fast sedimentation changing from 65-70 m.a to slow sedimentation:

The deposition of the continental sedimentation formations was relatively rapid:

The Cahuasas Formation was dated by Carrillo-Bravo [31] on the basis of its stratigraphic position to be younger than the Pliensbachian and older than the Callovian part of the Middle Jurassic.

Erben [35] proposes a Callovian age for this unit, while in the Sierra Madre Oriental Cantu-Chapa [30] assigns a Middle Callovian age for the Tepexic Formation.

Whereas, slow sedimentation formations are comprised of deep water depositional formations.

The Chipoco Formation according to Cantu-Chapa [28] interprets this unit as having a stratigraphic range from the Batonian to the Callovian.

Cantu-Chapa [30] assigns the Santiago Formation of a middle to late Callovian-Oxfordian age.

Magnetic anomaKas and positive gravity It infers, that the continental crust is thin compared to regional thicknesses.

The 1:50000 magnetic chart (Figure 38) of the Molango region shows a series of anomagnetic anomaKasmagnetics that stand out in the southwestern part of the chart where there are zones with an intensity greater than 350 Nanoteslas coexisting laterally with zones of -350 Nanoteslas. In the same way, a positive change in the intensity can be observed.

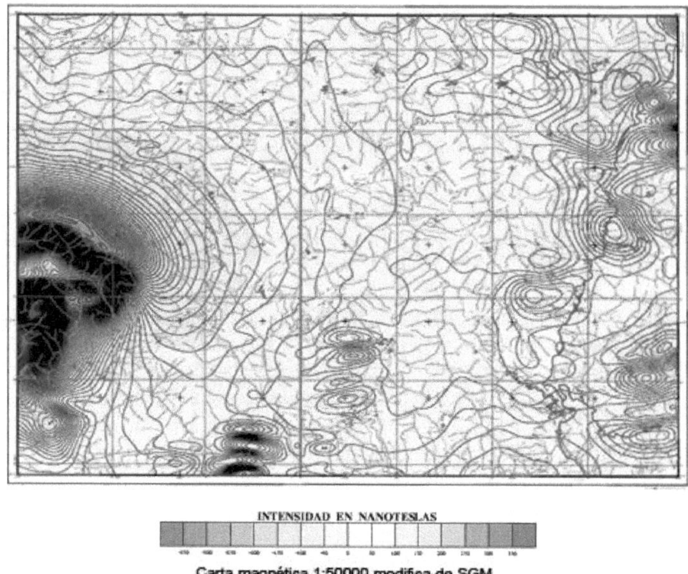

Figure 38, magnetics carrying a NW-SE direction.

3.1 Results of Geochemical Exploration Methods X-ray Diffraction (XRD)

Sample (F-HUAYA01) Huayacocotla Formation

Hand sample F-HUAYA01 was taken from an outcrop of the Huayacocotla Formation located at coordinates N 20°44.644' and W 98°42.566' at an altitude of 1524 masl. The sample has a light grey colour, sandy texture and a hardness of 7. The minerals found in this sample are distributed with more than 75% quartz, 10% feldspars, 5% lithic fragments and 20% minority minerals including pyrite, galena, chalcopyrite and jarosite. Therefore, according to the classification of Pettijohn *et al.*, 1973. It is classified as a quartzoarenite. Figure 39 shows the diffractogram of sample F-HUAYA01 where it was possible to identify the following minerals: Pyrite (99101-3104), High Quartz (99-200-3992), Alpha Quartz (01-089-8936), Albite (99-100-0593).

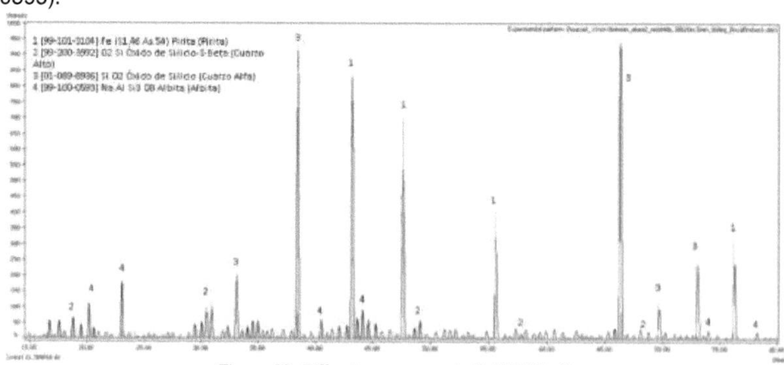

Figure 39. Diffractogram sample F-HUAYA 01

Sample (F-TEPEX01) Tepexic formation

Sample F-TEPEX01 was taken from the Tepexic Formation located at coordinates N 20°

444.977' W 98° 42.957' at 1530 m asl. It is a light grey calcarenite, weathered and dark grey when fresh, with a sandy texture that reacts with chlorhydric acid.

Figure 40 shows the diffractogram of sample F-TEPEX01 where it was possible to identify the following minerals: Calcite (01-086-2342), Montmorillonite (00-003-0010).

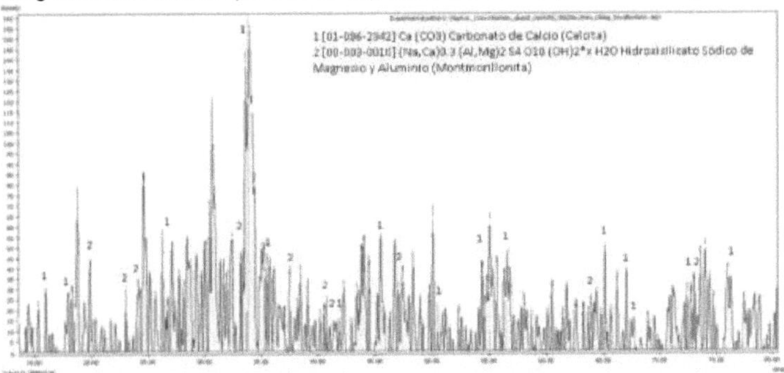

Figure 40. Diffractogram sample F-TEPEX01

Sample (F-SANT01) Formation Santiago

Sample F-SANT01 was taken at coordinates N 20°46.697' and W 98°43.563' at 1767 masl, outcrop of the Santiago Formation, it is a brownish-brown weathered and light grey fresh shale, the sample shows small fossils that were not possible to identify. Figure 41, represents diffractogram of sample F-SANT01 where it was possible to identify the following minerals: Hydroxyapatite (01-079-0685), Ankerite (01-083-15319), Dolomite (01-084-2024) and Calcite (00-003-0612).

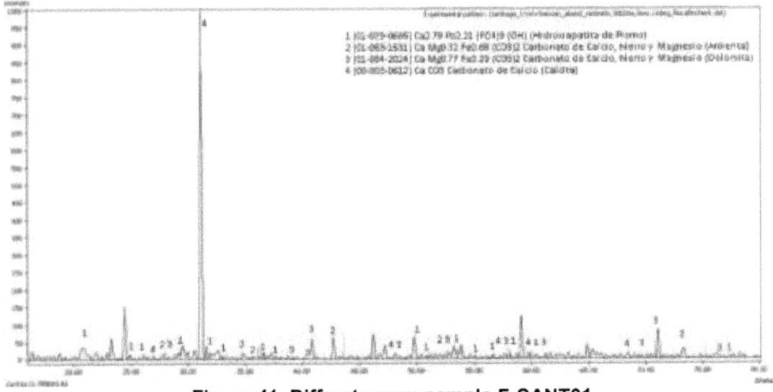

Figure 41. Diffractogram sample F-SANT01

Sample (F-CHIP01) Formacion Chipoco

Sample F-CHIP01 belongs to the Chipoco Formation and was taken from an outcrop located at coordinates N20°43.539' and W 98°42.982'. The sample is a dark limestone with reddish intemperate and blackish grey tones when fresh, it has a fine texture and reacts with chlorhydric acid. Figure 42 shows the diffractogram of the sample F-CHIP01 where it was possible to identify the following minerals: Manganese Oxides (99-001-1737), Rhodochrosite (00-001-0981) and Pyrrhotite (00-024-0220).

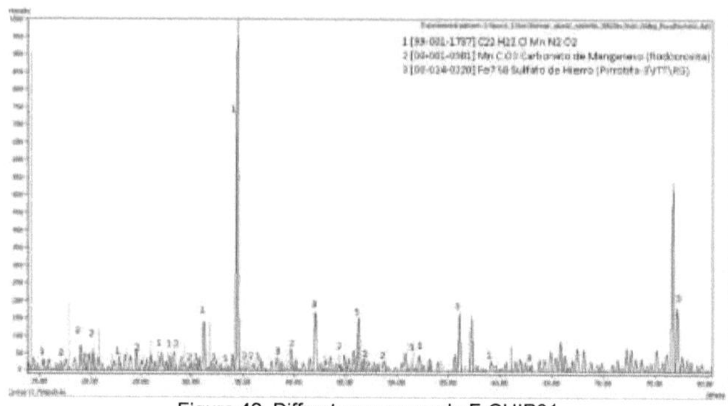

Figure 42. Diffractogram sample F-CHIP01

Sample (F-ATOTO01) Atotonilco Formation

Sample F-ATOTO01 was taken from an outcrop of the Atotonilco Formation, located at coordinates N 20° 20'11.2" and W 98°42'24.9" with an altitude of 2006 m.a.s.l., this sample is a light beige rhyolitic tuff. Figure 43 shows the diffractogram of sample F-ATOTO01, where it was possible to identify the following minerals: feldspar (99-100-1836), Albite (99-100-5774) and Quartz (00-003-0419).

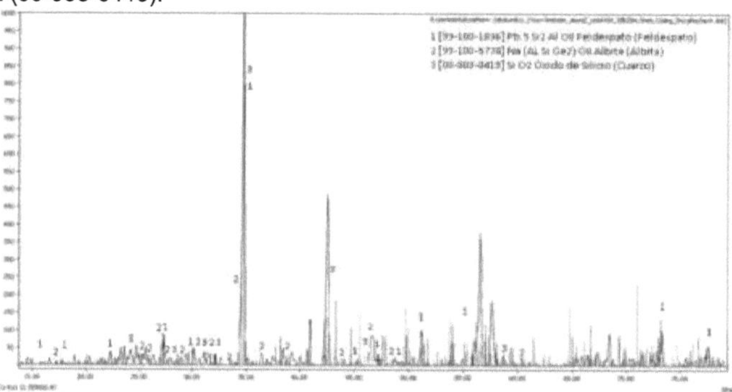

Figura 43 Difractograma muestra F-ATOTO01

The mineral phases found in the above samples provided evidence for the depositional environment as well as the basin conditions during the formation of the basin.

By means of XRD we were able to identify the relative abundance of at least 4 predominant minerals in the deposit studied through the drill core sample taken from the Huayacocotla formation. These are: Quartz (PDF: 00-046-1045), albite (00003-0508), calcium carbonate (01-087-1863), aluminium oxide (00-050-1496), and pyrite (01-076-0964).

Table 10. Mineral ratio and formation environments

		Mineral phases	Training environment	Training conditions
Atotonilco Formation	---	Feldspar (99-100-1036) Albite(99-100-5774) Quartz (30-003-0419)	Volcanic	Oxidant
Chipoco Training		Manganese oxides (99-001-1737) Rhodochrosite (00-001-0981) Pyrrhotite (00-024-0220)	Deep sea	Oxidant
Training Santiago	&	Ankerital01-083-15319) Dolomite (01-084-2024) Calcite (00-003-0612)	Quiet marine in basin facies	Gearbox

54

	┗ лч			
Tepexic Training	W ∎V.	Calcite (01-086-2342) Montmorllonite (00-003-0010)	Coastal to Deep	Oxidant
Huayacocotla Formation	5	Pyrite (99-101-31 04) High Quartz (99-200-3992) Alpha Quartz (01-089-8930) Albite (99-100-0593)	Deep to coastline	Gearbox

3.1 Bore core logging

A 9 metre long, 1.23 inch diameter core drill core, Figure 44, of quartz sandite was obtained using a Winkie Portable 2-cycle gas powered drill rig with air-cooled motor, and the first 3 metres were described every five centimetres for initial studies.

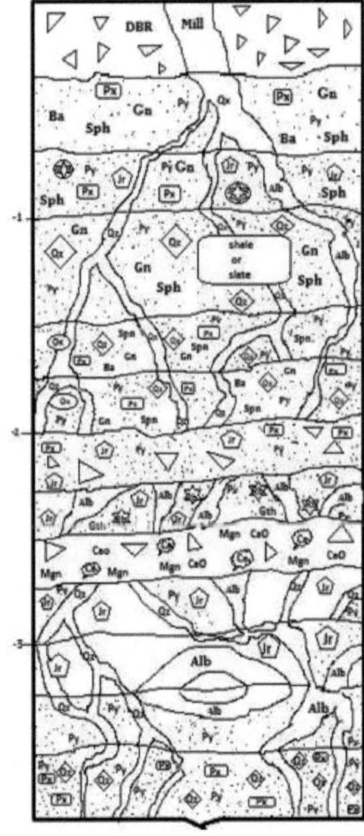

0-35cm. Zone of altered rock and brecciated material.

35-70cm. It presents zones of silicification, sulphides such as disseminated pyrite (Py), galena (Gn) and sphalerite (Sph),

70-100cm. This section presents druses that are constituted by minerals such as jarosite (Jr), quartz (Qz) and albite (Alb). It contains disseminated pyrite and

100-150cm. This zone presents sulphides such as pyrite and galena, quartz nodules, quartz and albite conduction channels, and a fragment of a quartz and an albite channel can be observed.

150-175cm. Minerals such as pyrite, galena, sphalerite, pyroxenes, as well as nodules and conduction channels of quartz and albite are observed.

175-200cm. This zone presents galena, pyrite and sphalerite, barite, pphroxenes, conduction channels and nodules of

200-225cm. Brecciated zone with the presence of minerals such as jarosite, pyroxenes and disseminated pyrite.

225-250cm. Presence of quartz conduction channels running through the whole section, with goethite, jarosite, rubidium and disseminated pyrite in the periphery.

250-275cm. Brecciated material showing caliche, talc, magnesite and calcium oxides.

275-300cm. Presence of quartz and albite conduction channels, with jarosite and pyrite in the periphery.

300-325cm. Along this zone there are several conduction channels and in the periphery there is pyrite.

325-350cm. This zone has quartz and albite veinlets around it, with disseminated pyrite.

350-375cm. Quartz and albite conduction channels are present, and minerals such as nodular quartz, disseminated pyrite and pyroxenes are found in the periphery.

Figure 44. Borehole description

3.1.1 Induced Coupled Plasma (ICP) Analysis Results

ICP analyses of samples collected from a drill core were interpreted to compare the elemental abundance between the mineralised zones delineated in this work as a prospecting point for a SEDEX deposit in the unmineralised Huayacocotla Formation and a normalised chondrite sample (UCC) (Taylor & McLeannan) Figure 45. The analysis yielded the following results: 82 ppm for Ba, 0.9 % Al, 0.17 % Ca, 1.64 % Fe, 0.08 % Ti, 40.8 % Si, Bi 2 ppm, 20 ppm Ce, 2.2 ppm Co, 30 ppm Cr, Cs 2.7 ppm, 0.9 ppm of Er, Ga 2.5 ppm, 1.6 ppm, 1.6 ppm of Er, Ga 2.5 ppm , 1.6 ppm Gd , Ge 1.5 ppm , 9 ppm of La, 71 ppm Li , 104 ppm Mn, 10 ppm Nd , 17 ppm Rb, 2ppm Se, Sr 9 ppm , 10 ppm Ta , Te 6 ppm , 1.3 ppm of U , V 28 ppm , 9 ppm , and 0.7 ppm Yb. While in lower proportion of precious metals found as Au < 0.02 g / t, Pd <0.05 g / t, Pt <0.05 g / t. It can be observed that the large-ion lyophilic elements (LILE) Cr = 30 and Th= 1.8 with respect to the parameters of the upper continental crust (UCC; Mclenan) show an enrichment. By presenting an increase in these elements we can infer a closeness to the rock coming from this detrital rock. On the other hand, the transition trace elements (TTE): V =28ppm, Cr =30 Cobalt =2.2 Cu = 7 Ni = 10 are present in a higher concentration than in the chondrite, which is associated with the felsic nature of the deposit and vanadium an affinity with elements of the platinum group.

Similarly, it shows high values for the lanthanide light rare earth elements (LREE) while the values for heavy rare earth elements (HREE) and thorium,Y =8:8 Gallium = 2.5 are below the Mclenan chondrite values.

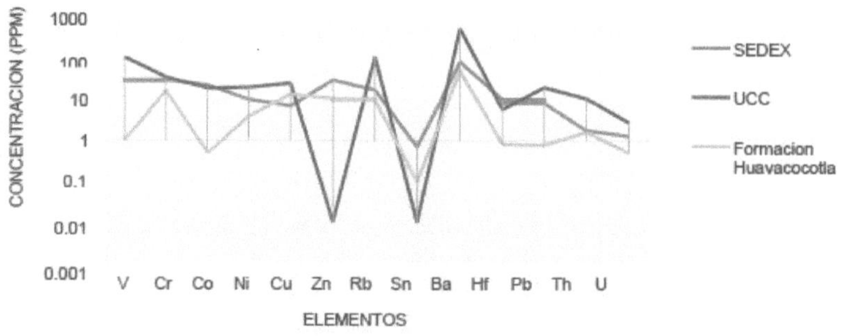

Figure 45. Table of UCC McLenan/Sedex/F.Huayacocotla element concentrations.

3.1.2 Scanning Electron Microscopy and Energy Dispersive Spectrometry (SEM-EDS) Results

For the scanning electron microscopy, a representative sample of two metres (huaya_2) was taken to observe the type of textures that are generally present in the sample, with which the following was described: in Figure 46, different microstructures are shown, where three representative zones are observed A (pyrite framboid), B (clay minerals), C (albite twins).

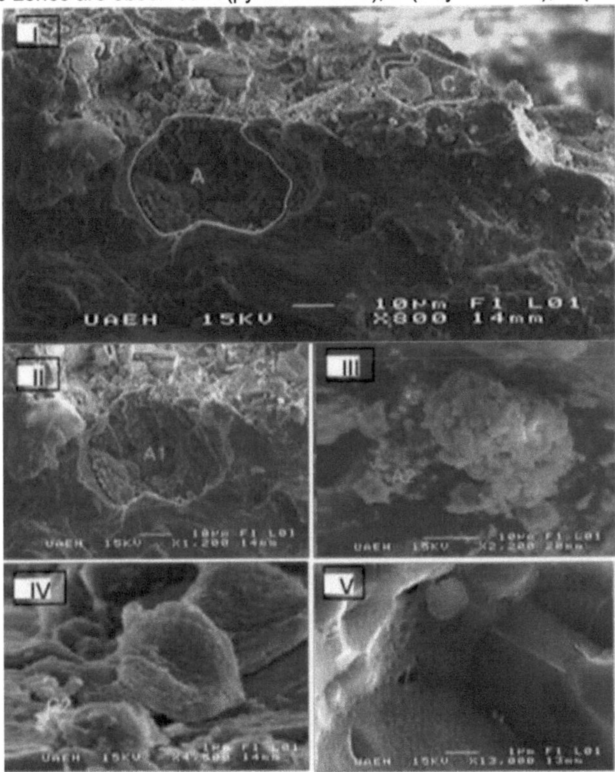

Figure 46, microstructures of a mineral SEDEX

In them it is possible to observe in I parUculas that are presented in the order of 10 pm in which anhedrales partfculas are distinguished. II, an approach of a remnant of a framboidal structure; in III a framboidal structure of pyrite, possibly caused by bacterial activity, in IV an approach of a framboid of pyrite, and in 3 a twinned structure of albite in the centre.

In the following diagrams, figures 47, 48 and 49, the interaction of the elements gold, and silver in relation to the elements copper, lead and platinum respectively is represented, representing the distribution at depths of one metre, two metres and three metres respectively.

In the ternary diagram Au+Ag+Cu, it can be observed that for a depth of one metre represented by the blue vane the element with the highest content is gold, however, in the sample taken at a depth of two metres represented by the green vane, silver is found in higher proportion, later for the sample taken at a depth of three metres represented by the red vane, the ppm levels of copper decrease to 0.06.

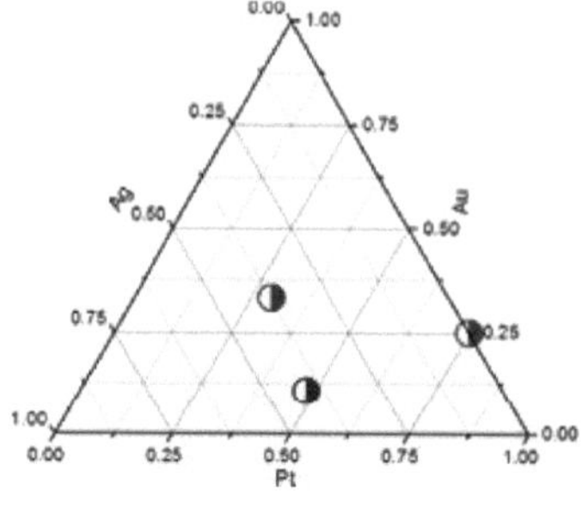
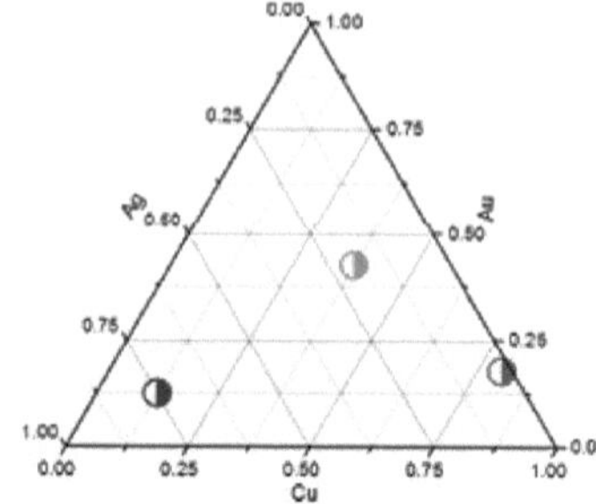

Figure 47. Au+Ag+Cu ternary diagram. Figure 48. Au+Ag+Pt ternary diagram.

In the ternary diagram Au+Ag+Pt it can be seen that for the one metre depth represented by the cherry vine and the two metre depth represented by the blue vine, the Au content predominates, while for the three metre depth represented by the purple vine, the Au content predominates, while for the three metre depth represented by the purple vine the Au content predominates.

In the ternary diagram Au+Ag+Pb it can be seen that at one metre depth represented by the green vignette the sample shows a higher abundance of Au, on the other hand, the sample at two metres depth represented by the blue vignette shows a predominance of Au.

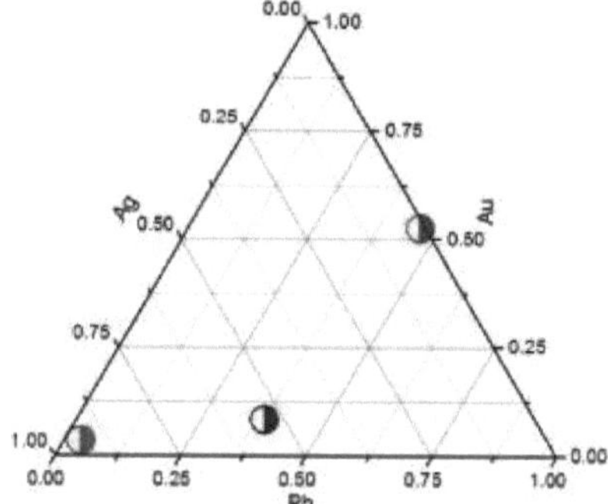

Figure 49. Ternary diagram Au+Ag+Pb Ag content, while for the three-metre sample represented by the cherry vineyard the major element is Pb.

In the diagram Al2O3+Fe2O3+MgO figure 50, the red vignette represents the sample taken at one metre and where the predominance of Al2O3 can be observed as well as the sample at two metres.

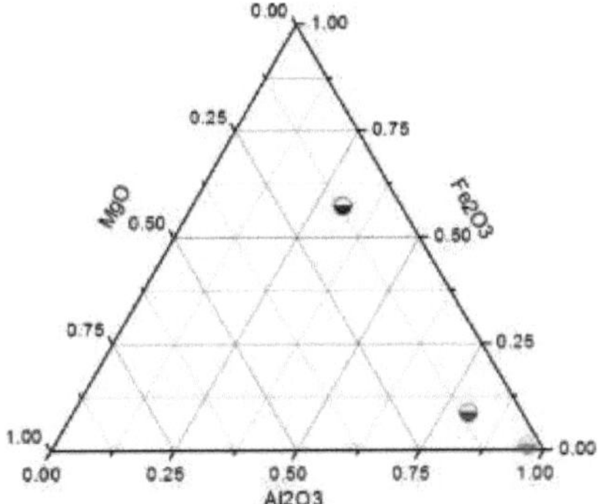

Figure 50. Ternary diagram Al2O3+Fe2O3+MgO

in green on the other hand, the sample at three metres in blue is dominated by Fe2O3 .
In the diagram Al2O3+Fe2O3+CaO Figure 51, at the first metre represented by the red vignette, at the two metres represented by the green vignette there is a predominance of Al2O3 while, for the three metres represented by the blue vignette, CaO is found to be higher.

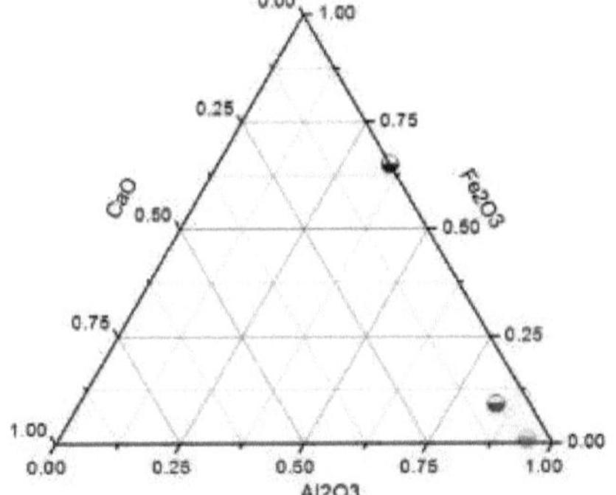

Figure 51. Diagram Al2O3+Fe2O3+CaO

Finally, in the diagram Al2O3+Fe2O3+SiO2 Figure 52, it can be observed that at the depths the samples show a tendency towards SiO2

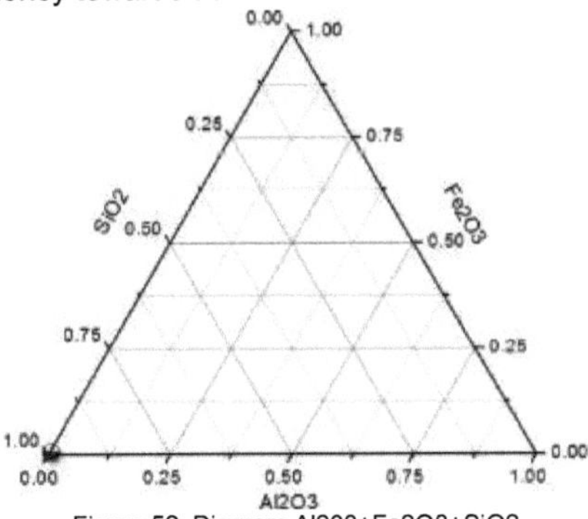

Figure 52. Diagram Al2O3+Fe2O3+SiO2

59

DISCUSSION

Indirect method of mineral exploration: The mining exploration method proposed here, is focused on Prospecting for mineralisation in sedimentary rocks whose minerals are preferentially hosted in strata or related to these by stratigraphic relationships of marine transgressions, over rocks of continental origin and thus^ find and show susceptible traces that could contain mineralisation in a given area. Although its application here relates only to sedimentary rocks, it could also be used indirectly to define other lithologies such as in ^gneissic and metamorphic rocks, which may contain mineralisation. Furthermore, it should be noted that sedimentary deposits are widespread and occur in all geological eras, from the Proterozoic to the present day, and in different depths. On the other hand, secular variations in the abundance of sedimentary deposits occur, for example in the study area. during the Lower Jurassic, but the abundance of occurrences could also be determined by correlating with other major mineral deposits already in exploitation.

Characterisation of the ore deposit: In the geochemical analysis of trace elements such as Cr and Th, due to their low mobility, these are considered suitable for determining provenance and tectonic setting [30]. As can be seen from the results, there is an enrichment of large ion elements such as Cr=80ppm and Th=25ppm; all are higher than the average content of an arid Huayacocotla shale. On the other hand, the contents (in ppm) of transition trace elements V=89, Co=15.8, Cu=331 and Ni=41.5, are low because the affinity of the deposit could be either felsic or acidic. Indeed, while the Cr and Ni content is low, the V content could be associated with the occurrence of metals such as platinum, palladium and gold. Furthermore, this deposit has a higher concentration of light rare earths (LREE) than the Clarke, because the values (in ppm) for heavy crude earths (HREE) are low for Y=32.3 and Ga=21. For this reason, the presence of associated REE minerals such as monazite (Ce, La, Nd, Th, (PO4)) and bastnasite could be inferred. It is important to note that the negative HREE anomaKa for Eu and Hf with respect to Clarke confirms the HREE depletion and reaffirms the felsic affinity of the deposit, and the possibility that its origin is from a gneissic protoKtic, because La mineralization has higher concentrations of LREE. Similarly, a negative Ta anomaKa shows a sedimentary affinity of the mineral deposit. On the other hand, the negative sulphur anomaKa may be due to the presence of pyrite formed by sulphate reduction in the marine environment. lower Jurassic age environment. In addition, the possible organic carbon contents, the presence of pyrite in aggregate botroiidal sulphur clasts and low base metal contents, low Ag/Au ratios, and minor arsenopyrite and pyrrhotite contents, could be to the silistlastic nature that *is typical of Rift environments.*

Likewise, the study area according to the results obtained could be classified as a rift, as described by Jowet, showing in addition to bimodal volcanism, fault-controlled patterns, positive gravity and rapid sedimentation rates over short periods of time that change at slow rates over long periods. Finally, the Huayacocotla Formation in which this deposit is found, has a great similarity in lithological, stratigraphic and chronological aspects with another deposit found in the San Cayetano Formation, which was also classified as SEDEX and is located in the Matahambres area of Cuba.

CONCLUSIONS

I.   An innovative method of Mineral Exploration was developed and evaluated through field work, and with which sedimentary deposits were found in the eastern part of the state of Hidalgo, Mexico. This method, based on the study of transgressions, has been named "Metodo Indirecto de Exploration Minera".

II.   The mineralisation found is strongly linked to a sedimentary control; it is also closely related to a marine transgression of Lower Jurassic age.

III.   The mineral outcrops encountered consist of two types of mineralisation. The first is of the phyllonian type, where the exhalation. Stock-work type rakes were observed at the base. And the second, of stratiform type formed by a sequence of mutually concordant sandstones and shales of marine origin. This preliminary lithology is the Hpico found in several SEDEX formations.

IV.   According to the characterisation by thin sections, polished sections, metamorphism and cataclastic fragments, pyrite and quartz were found to be mechanically deformed, which is indicative of the existence of ancient tectonism and the movement of transgressions. Likewise, minerals from a reducing environment, such as pyrite, were found in botryoidal form, indicating its marine origin. In the same way, it was possible to identify micro-veins full of quartz and disseminated pyrite, remobilisation minerals such as chalcopyrite and finally monazite; all of them possibly originated in a stock-work.

iv. From the analysis performed by ICP, as^ well as from the complementary characterisation performed by XRD, SEM-EDS; it is concluded that the ore found contains adequate values of precious metals such as Au, Ag, Pt and Pd, as^ well as some rare earth elements, since their values are above the average classification made by Clarke. vi. All of the above contributes to establish that 86 depending on the results obtained; this outcrop is a SEDEX type deposit with potential mining potential. Furthermore, these results validate the application of the Indirect Mineral Exploration Method to find sedimentary deposits through the study of transgressions, but could be modified to find any type of deposit, depending on its main lithological characteristics.

REFERENCES

1. WaliArain, A.Y.; ShakoorMastoi, A.; DaaharHakro, A.A.A.; AhmedRajper, R.; AfzalJamali, M.; RazaBahatti, G.; Bhatti, W. A preliminary review on the metallogeny of sediment-hosted Pb-Zn deposits in Balochistan, Pakistan. Earth Sci. Malasya, 2021, 5(1), 19-26. DOI: 10.26480/esmy.01.2021.19.26

2. Ishihara, S.; Kanehira, K.; Sasaki, A.; Sato, T.; Shimazaki, Y. Geology of Kuroko deposits. Min. Geol., 1974, special issue, volume 6, pp. 435.

3. Eldridge, C.S.; Barton, P.B. & Ohmoto, H. Mineral textures and their bearing on formation of the Kuroko ore bodies. Pub.geoscienceworld.org, 1983, DOI: 10.5382/Mono.05.15

4. Santoro, L.; Putzolu, F.; Mondillo, N.; Boni, M. and Herrington, R. Influence of Genetic Processes on Geochemistry of Fe-oxy-hydroxides in Supergene Zn Non-Sulfide Deposits. Minerals, 2020, 10(7), 602. DOI: 10.3390/min10070602

5. Castillo-Oliver, M.; Melgarejo, J.C.; Torro, L.; Villanova-de-Benavent, C.; Campeny, M.; D^az-Acha, Y.; Amores-Casals, S.; Xu, J.; Proenza, J. and Tualer, E. Sandstone-Hosted Uranium Deposits as a Possible Source for Critical Elements: The Eureka Mine Case, Castell-Estao, Catalonia. Minerals, 2020, 10(1), 34. DOI: 10.3390/min10010034

6. Jowett, E.C. Effects of continental rifting on the location and genesis of stratiform copper-silver deposits. Sediment-hosted stratiform copper deposits. Geological Association of Canada, Special Paper, 1989, 36, 53-66.

7. Dostal, J. and Gerel, O. Occurrence of Niobium and Tantalum Mineralization in Mongolia. Minerals, 2022, 12(12), 1529. DOI: 10.3390/min12121529

8. Balaram, V. Potential Future Alternative Resources for Rare Earth Elements: Opportunities and Challenges. Minerals, 2023, 13(3), 425. DOI: 10.3390/min13030425.

9. Liu, X.; Yang, K.; Rusk, B.; Qiu, Z.; Hu, F. and Pironon, J. Copper Sulfide Remobilization and Mineralization during Paleo-protezoric Retrograde Metamorphism in the Tongkuangyu Copper Deposits, North China Craton. Minerals, 2019, 9(7), 443. DOI: 10.3390/min9070443

10. Wang, J.; Wang, X.; Liu, J.; Liu, Z.; Zhai, D. and Wang, Y. Geology, Geochemistry, and Geochronology of Gabbro from the Haoyaoerhudong Gold Deposit, Northern Margin of the North China Craton. Minerals, 2019, 9(1), 63. DOI: 10.3390/min9010063

11. Kui-Feng, Y.; Hong-Rui, F.; Pirajno, F.; Liu, X. Magnesium isotope fractionation in differentiation of mafic-alkaline-carbonatitic magma and Fe-P-REE-rich melt at Bayan Obo, China. Ore Geo. Rev., 2023, volume 157, 105466. DOI: 10.1016/j.oregeorev.2023.105466

12. Obaidalla, N.A.; Mahfouz, K.H. & Metwally, A.A. Mesozoic Sedimentary Succession in Egypt. In: The Phanerozoic Geology and Natural Resources of Egypt. Cham: Springer International Publishing, 2023, p. 169-219.

13. Ogden, C.S.; Bastow, I.D.; Ebinger, C.; Ayele, A.; Kounoudis, R.; Musila, M.; Bendick, R.; Mariita, N.; Kianji, G.; Roonoy, T.O.; Sullivan, C.; Kibret, B. The development of multiple phases of superposed rifting in the Turkana Depression, East Africa: Evidence from receiver functions. Earth Plan. Sci. Letters, 2023, volume 609, 118088. DOI: 10.1016/j.epsl.2023.118088

14. Jara, R.E.G.; Ghiglione, M.C.; Galliani, L.R.; & Mpodozis, C. From rift to foreland basin: A case example from the Magal-lanes-Austral basin, southernmost Andes. Basin Research, 2023, 35(3), 865-897. DOI: 10.1111/bre.12739

15. Sillitoe, R.H. & Rodriguez, G. Exhalative red-bed copper mineralization in travertine, Puna Plateau, northwest Argentina. Miner Deposita, 2023, 58(2), 243-261. DOI: 10.1007/s00126-022-01134-y

16. Walter, B.F.; Giebel, R.J.; Siegfried, P.; Doggart, S.; Macey, P.; Schiebel, D.; Kolb, J. The genesis of hydrothermal veins in the Aukam valley SW- A far field consequence of

Pangean rifting? J. Geochem. Explo. 2023, volume 250, 107229. DOI: 10.1016/j.gexplo.2023.107229

17. Lisboa, L.H.D.; Filho, C.F.F.; Monteiro, L.V.S. & Mansur, E.T. Gahnite, garnet and manganite compositions of metamor-phosed sediment-hosted Zn-Pb-(CuAg) deposits of the Mesoproterozoic Nova Brasilandia Group: Vectors for SEDEX de-posits with Broken Hill-type affinities in the western Amazonian Craton, Brazil. J. Geochem. Explor., 2023, volume 249, 107210. DOI: 10.1016/j.gexplo.2023.107210

18. Cerecedo-Saenz, E.; Rodriguez-Lugo, V.; Hernandez-Avila, J.; Mendoza-Anaya, D.; Reyes-Valderrama, M.I.; Moreno-Perez, E. and Salinas-Rodriguez, E. Mineralization of Rare Earths, Platinum and Gold in a Sedimentary Deposit, Found Using an Indirect Method of Exploration. Aspects Min Miner Sci, 2018, 1(2), 1-9. DOI: 10.31031/AMMS.2018.01.000510

19. Perez-Vazquez, R.G.; Melgarejo i Draper, J.C. The Matahambres deposit (Pinar del Rio, Cuba): structure and mineralog^a (In Spanish). Acta geologica hispanica (In Spanish), 1998, Vol. 33, Num. 1, pp. 133-152. https://raco.cat/index.php/ActaGelogica/article/view/75549

20. Okita, P.M. Manganese carbonate mineralization in the Molango District, Mexico. Econ Geo, 1992, 87(5): 1345-1366. DOI: 10.2113/gsecongeo.87.5.1345

21. Wang, J.; Xuexiang, G.; Jinchi, X.; Yongmei, Z.; Yiwei, P.; Liangtao, L. New Insights into the Tectonic Setting and Origin of the Haerdaban Pb-Zn Deposit. Chinese Western Tianshan: Evidence from Geological, Chert Geochemistry, and Detrital Zircon U-Pb Geochronology. Posted 13 Jun 2023, 57 pages, Available at SSRN: https://ssrn.com/abstract=4477187 or https://dx.doi.org/10.2139/ssrn.4477187

22. Cawood, T.K.; Rozendaal, A. & Spry, P.G. Discussion on "Syn-metamorphic sulfidation of the Gamsberg zinc deposit, South Africa" by Stefan Hohn, Hartwig E. Frimmel, and Westley Price. Miner Pretol, 2023, DOI: 10.1007/s00710-023-00821-6

23. Creus, P.K.; Sanislav, I.V.; Dirks, P.H.G.M.; Jago, C.M.; Davis, B.K. The Dugald Rivertype, shear zone hosted, Zn-Pb-Ag mineralisation, Mount Isa Inlier, Australia. Ore Geo Rev, 2023, volume 155, 105369. DOI: 10.1016/j.oregeorev.2023.105369

24. Liu, W.; Mei, Y.; Etschmann, B.; Glenn, M.; MacRae, C.M.; Spinks, S.C.; Ryan, C.G.; Brugger, J.; Paterson, D.J. Germanium speciation in experimental and natural sphalerite: Implications for critical metal enrichment in hydrothermal Zn-Pb ores. Geo Cosmochem Acta, 2023, volume 342. pp. 198-214. DOI: 10.1016/j.cga.2022.11.031

25. Conde, C.; Tornos, F.; Danyushevsky, L.V. & Large, R. Laser ablation-ICPMS analysis of trace elements in pyrite from the Tharsis massive sulphide deposit, Iberian Pyrite Belt (Spain). J Iberian Geo, 2021, 47, pp. 429-440. DOI: 10.1007/s41513-020-00161-w

26. Williams, N. Light-Elements Stable Isotope Studies of the Clastic-Dominated Lead-Zinc Mineral Systems of Northern Australia and the North American Cordillera: Implications for Ore Genesis and Exploration. In: Huston, D.; Gutzmer, J. (eds) Isotopes in Economic Geology, Metallogenesis and Exploration. Mineral Resource Reviews. Springer, Cham. DOI: 10.1007/978-3-031-27897-6_11

27. Staude, S.; Raisch, D. & Markl, G. Sulfide anatexis during high-grade metamorphism: a case study from the Bodenmais SEDEX deposit, Germany. Miner Deposita, 2023, 58. pp. 987-1003. DOI: 10.1007/s00126-023-01166-yAguayo-Camargo, J. E. (1977). Sedimentation and diagenesis of the Chipoco Formation (Upper Jurassic) in outcrops, Hidalgo and San Luis Potosi States. *Journal of the Mexican Petroleum Institute*, 11-37.

Barry Maynard, J., & D. Klein, G. (1993). Tectonic Dubsidence Analyses in the Characterization of Sedimentary Ore Deposits: Examples From the Witwatersrand (Au), White Pine (Cu), and Molango (Mn). *Economic Geology* , 37-50.

28. Cantu-Chapa, A. (1969). Stratigraphy of the Middle-Upper Jurassic of the Poza Rica

Subsoil; Ver. *Journal of the Mexican Petroleum Institute, v. 1* , 3-9.

29. Cantu, Chapa, A., 1971, La Serie Huasteca (Jurasico Medio-Superior) del Centro Este de Mexico: Revista del Instituto Mexicano del Petroleo, 3, 17-40.

30. Cantu-Chapa, A. (2001). Mexico as the Western Margin of Pangea based on Biogeographic evidence from the Permian to the Lower Jurassic. *AAPG* , 1-27.

31. Carrillo-Bravo, J. (1965). Geological study of a part of the Huayacocotla anticlinorium. *Boletin de la Asociacion Mexicana de Geologos Petroleros, v.17,* 73-96.

32. Cerecedo Saenz, E. (2003). Cu and Ag mineralization of the Triassic-Jurassic Rift of eastern Mexico.

33. Cerecedo Saenz, E., Salinas R., E., Cantu C., A., Patino C., F., Ramirez C., M., HernandezA., J., et al. (2009). The Chipoco Formation and its stratigraphic relationship with the possible manganeseiferous deposits of Chichapala; Veracruz. *Revista de investigacion cientifica del CIMMMySH* , 40-52.

34. Ochoa-Camarillo, H.R., 1997, Aspectos bioestratigraficos, paleoecologicos y tectonicos del Jurasico (anticlinorio de Huayacocotla) en la region de Molango, Hidalgo, in Gomez-Caballero, A., Alcayde-Orraca, M. (eds.), II Convencion sobre la Evolucion Geologica de Mexico y Recursos Asociados, Pachuca, Hidalgo, Mexico: Mexico, Universidad Autonoma del Estado de Hidalgo, Instituto de Investigaciones en Ciencias de la Tierra, UAEH, Instituto de Geolog^a, UNAM, Symposium and Colloquium, unpaginated.

35. Erben, H. (1956). The Lower Jurassic of Mexico and its ammonites. *International Geological Congress* (p. 393). Mexico, D.F.: Monografia.

36. Imlay, R.W, 1953, The Jurassic Formations of Mexico: Boletm de la Sociedad Geologica de Mexico, 16, 1-65.

37. Hermoso-de la Torre, C., & Martinez-Perez, J. (1972). Detailed measurement of Upper Jurassic formations in the Sierra Madre Oriental Front. *Asociacion Mexicana de Geologos Petroleros,* 45-64.

38. Robin, C., 1976b, Las series volcanicas de la Sierra Madre Oriental (basaltos e ignombritas) description y caracteres quimicos: Revista Instituto de Geolog^a, Universidad Nacional Autonoma de Mexico, 2, 12-42.

39. Abelardo Cantu Chapa (1998), Las transgresiones Jurasicas de Mexico, Revista mexicana de ciencias geologicas, volumen 15, numero1 , 1998,p 25-37, Universidad Autonoma de Mexico, Instituto de Geologia y Sociedad Geologica Mexicana, Mexico DF.

40. Kirkham, R.V., 1989, Distribution, setting and genesis of sediment-hosted stratiform copper deposits, in Boyle, R.W., Brown, A.C., Jefferson, C.W., Jowett, E.C., Kirkham, R.V. (eds.), Sediment-hosted stratiform copper deposits: Geological Association of Canada, Special Paper 36, 3-38.

41. Goodfellow, W.D., Lydon, J.W. (2007) Sedimentary exhalative (SEDEX) deposits. In: Goodfellow, W.D. (Ed.) Mineral deposits of Canada: a synthesis of major deposit types, district metallogeny, the evolution of geological provinces, and exploration methods. Geological Association of Canada Special Publication 5, 163-183.

42. Peter Bartok (1993) Prebreakup geology of the Gulf of Mexico-Caribbean: Its relation to Triassic and Jurassic rift systems of the region. P Bartok. Tectonics 12 (2), 441-45

I want morebooks!

Buy your books fast and straightforward online - at one of world's fastest growing online book stores! Environmentally sound due to Print-on-Demand technologies.

Buy your books online at
www.morebooks.shop

Kaufen Sie Ihre Bücher schnell und unkompliziert online – auf einer der am schnellsten wachsenden Buchhandelsplattformen weltweit! Dank Print-On-Demand umwelt- und ressourcenschonend produziert.

Bücher schneller online kaufen
www.morebooks.shop

Printed by Books on Demand GmbH, Norderstedt / Germany